Climate Change and Clean Energy Management

Climate change has never been more important than it is now, as it has become arguably the world's most urgent problem. Solving this problem is proving difficult and complex as it involves joint efforts by governments, companies, communities and innovators. The increased use of fossil fuels associated with global economic growths has led to rising GHG emissions and global warming. There are many challenges for countries that are enacting new climate and clean energy regulations in line with their Paris Agreement commitments.

Good government policies and corporate strategies are essential to support these efforts as part of the global climate change crisis. This important book addresses the latest climate change impacts and developments in potential mitigation strategies. These include fossil to clean energy transition, smart low-carbon city designs, green transportation, electric vehicles, green agriculture, carbon emission trading, carbon capture solutions plus climate finance and risk management. Potential new policies and strategies to support the successful implementation of these important strategic areas are discussed together with high level country and business case examples.

This book is essential reading for policy makers, government employees, business executives, professionals, researchers and academics alike looking to affect change to global climate and energy policies.

Henry K. H. Wang is an international adviser, author and speaker with extensive high-level business experience. He is President of Gate International and a former director of both Shell China and SABIC in Riyadh. He is a Fellow of the Royal Society of Arts FRSA and Fellow of the Institution of Chemical Engineers. He has been invited to join the London University SOAS Advisory Board and Imperial College Grantham Institute Climate Change Stakeholder Committee plus China Carbon Forum. He has published three books as well as over 100 papers and speeches globally. He has been invited to speak at international conferences, leading universities and business schools around the world.

Climate Change and Clean Energy Management

Challenges and Growth Strategies

Henry K. H. Wang

Routledge
Taylor & Francis Group

LONDON AND NEW YORK

First published 2020
by Routledge
2 Park Square, Milton Park, Abingdon, Oxon OX14 4RN

and by Routledge
52 Vanderbilt Avenue, New York, NY 10017

Routledge is an imprint of the Taylor & Francis Group, an informa business

© 2020 Henry K. H Wang

British Library Cataloguing-in-Publication Data
A catalogue record for this book is available from the British Library

Library of Congress Cataloging-in-Publication Data
A catalog record has been requested for this book

ISBN: 978-1-138-48488-7 (hbk)
ISBN: 978-1-351-05071-5 (ebk)

Typeset in Times New Roman
by codeMantra

Contents

Author's notes

This book is based on the author's research, literature surveys, high-level business experience and learning accumulated over some 40 years of successful international business globally. He has worked as senior executive and international adviser, author and speaker and as a director and board member of leading companies. He has also been invited to advise various leading universities, international institutions and companies.

The views expressed in this book represent the author's contribution as part of his global corporate social responsibilities, whilst also complementing his support for the development of future thought leaders and outstanding young people from around the world. It is hoped that this book will be of help to executives and professional practitioners as well as academics, researchers and students.

About the author

Henry K. H. Wang is an international adviser, author and speaker with extensive high-level business experience globally. He is President of Gate International Ltd. He is a former director of Shell China and SABIC in Riyadh. He has held various board roles in public companies and Joint Ventures world-wide. He has been invited to advise leading universities, international institutions and companies and he has also been invited to speak regularly at international conferences and give media interviews. Leading universities and business schools frequently invite him to speak and lecture.

He is a Fellow of the Royal Society of Arts FRSA and a Fellow of the Institute of Chemical Engineering, UK. He has been invited to join the London University SOAS SCI Advisory Board and the University of London China Advisory Board. He has also been invited to join the UK Climate Change Committee Working Group and the Imperial College London Grantham Institute of Climate Change Stakeholder Committee plus the China Carbon Forum Advisory Board. He was a former Vice Chairman of the OECD Business Energy & Environment Committee, plus former Vice Presidents of the EU & British Chambers of Commerce of China. He is a graduate of Imperial College London and University College London. He has also undertaken advanced management courses at Wharton and Tsinghua.

He has published books and technical and management papers globally and has also held various international patents on new inventions. His first book, *Successful Business Dealings and Management with China Oil, Gas and Chemical Giants*, was published in 2014. His second book, *Energy Markets in Emerging Economies: Strategies for Growth*, was published in 2017 and his third book, *Business Negotiations in China: Strategy, Planning and Management*, was published in 2018. His negotiation management paper was selected as one of the Top Five UK Management Papers of the Year in 2015 and published by UK Chartered Management Institute globally.

Preface

This new book aims to provide a holistic overview of the institutional, organisational and management issues that underpin successful climate change and clean energy management and growth, with international case studies.

Climate change and clean energy management have been receiving a lot of attention globally. Industrialisation and rising fossil fuels consumption globally have resulted in rising emissions and pollution. These have led to rising global warming and worsening environmental pollution. There is currently significant international pressure and drives to decrease greenhouse gas emissions and to reduce global environmental pollution. Many countries are promoting clean energy usage and reducing fossil fuel consumption so as to reduce their carbon emissions and environmental pollution. Looking ahead, climate change management and clean energy growth are expected to increase as major countries continue to push ahead with environmental improvements and to meet their Paris Agreement commitments. There are also many serious challenges to future climate change management and clean energy growth globally, particularly in strategic management, innovation and fierce competition. Good government policies and corporate strategies will be essential to support future successful climate change management and clean energy growth.

This book will address the latest developments in climate change and clean energy management, together with the key strategic challenges and risks. Potential new policies and strategies for climate change management and the future growth of clean energies, together with relevant high level business case examples, will be discussed.

The book should be of interest to a wide range of readers from different market segments, including professionals, practitioners, executives, academics and researchers globally who are actively involved with climate change and clean energy management, study and research.

The book should be attractive to professional readers, including policy makers, regulators and government officials, who have responsibilities for developing climate change policies and regulating clean energies globally. The book should also be of interest to business executives and practitioners, with its practical and commercial perspectives on the business contexts on climate change management. The new book should also help academics, researchers and students to support their studies and research into climate change and clean energy management theories and practices of both a situational and contextual nature. The university courses that the new book could be used for include climate change, environment, finance, international business and strategy. The book will also be relevant for new executive management courses on climate change, clean energy and climate finance management which have recently been introduced by leading business schools around the world.

This book complements the three Taylor & Francis books already published by the author globally, *Successful Business Dealings and Management with China Oil, Gas and Chemical Giants*, published in 2014, *Energy Markets in Emerging Economies: Strategies for Growth*, published in 2017 and *Business Negotiations in China: Strategy, Planning and Management*, published in 2018. These books have been included by leading universities on their reference lists for student courses and have been frequently cited by universities and researchers world-wide.

Acknowledgements

I would like to acknowledge the valuable input, support and encouragement that I have received from many senior executives, thought leaders and key stakeholders in top universities and business schools globally whilst working on this book.

My sincere thanks to Taylor & Francis, who commissioned and published this book. I would like to thank all the editorial and production staff who have contributed to the successful design, editing, copywriting, typesetting, proof-reading, publication and marketing of this book.

I would also like to thank the leading universities and institutes around the world who have invited me to advise, speak and work with them on climate change and clean energy. The keen interest from the university staff and students plus the professional practitioners has motivated me to write this book so as to share my experience and research with professionals, executives, academics and students globally.

I would like to sincerely thank my late wife and two wonderful children for all their great support, love, understanding and encouragement, which are much appreciated and treasured every day. I would also like to thank my Mother, sisters and their families plus our close friends for their valuable advice and support. Their strong support and encouragement have been essential to keep me going to complete the book, with the large amounts of personal time and effort required for the extensive research, writing and editing.

I am dedicating my fourth book to my dear late wife, whom I am missing dearly and treasuring all my fond memories of her.

1 Climate change global developments and impacts

十年树木,百年树人
shí nián shù mù, bǎi nián shù rén
Ten years for a sapling to grow into a tree and a hundred years to develop enterprises.
Good wine takes time to mature.

Executive overview

Climate change and clean energy management are receiving a lot of attention internationally. Industrialisation and rising fossil fuels consumptions globally have led to increased pollutions and carbon emissions. These have led to rising global warming and serious environmental pollutions in different countries. There are rising frequencies of climate-induced extreme weather incidents, including extreme weather, higher rainfall, rising sea levels, serious droughts and flooding, etc. These climate-induced incidents have caused serious disruptions and damages to various cities and countries globally. The various climate change impacts have raised serious challenges and risks to the continued sustainable economic growths and developments of key developed and emerging economies globally. The serious climate change impacts and challenges globally will be discussed in this chapter, with international examples.

Climate change and environment definitions

Climate change and clean energy management are key areas that have been receiving a lot of attention globally from scientists, companies and stakeholders. Literature surveys and research have highlighted that there are many different ways of defining climate change and clean energy management. These are often related to different ways of explaining what they are and their key implications globally.

Climate change is usually defined as the significant, long term changes in the global climate conditions, such as temperature or rainfalls, in different regions on earth. Climate changes have been usually caused by major, long term variations in the key global weather conditions. The changes in global and regional climate patterns have been particularly apparent and important from the mid-20th century onwards. Climate changes have also been largely attributed to the rise of global warming and global greenhouse effects (NASA, 2014).

It is important to understand the differences between weather and climate. Weather is generally referred to as short-term changes which involve various weather conditions such as rainfall, temperatures, humidity plus wind speeds and directions for a specific region or a city. Weather conditions could vary from hour to hour and day to day. The climate of a region or city is normally its weather conditions averaged over many years.

The climate of a city or region on earth will generally change more slowly. Climate changes have often led to changes in the typical weather of a region over a long period. A good example is the climate-induced changes to a region's averaged annual rainfalls over seasons. Climate change is also the changes in the earth's overall climate conditions. Good examples would be the changes in the earth's average temperatures and typical precipitation patterns. The global climate is usually more than the average of the climates of specific cities or places. The systematic connectivity of the global climate systems has contributed to climate change impacts being felt globally. Climate changes have led to higher global temperatures and global warming. These have then led to more frequent climate- induced extreme weather incidents. These have resulted in significant impacts on the daily life of populations around the world.

Global warming is usually defined as the rise of global temperatures resulting from the global greenhouse effects. The global greenhouse effect is usually defined as the process whereby the major greenhouse gases (GHG), such as water vapour, carbon dioxide (CO_2) and methane, have been impacting the earth's climate and environment. These GHGs have been generated by various human activities, such as industry, agriculture and transport. GHG emissions into the earth's atmosphere have caused damages to the ozone layer. They have also been absorbing or re-emitting the heat being radiated from the sun and various regions on earth. These have then resulted in the trapping of various heat emissions, leading to rising global temperatures and global warming around the world (NRDC, 2016).

Global warming has resulted in slow increases in the average temperature of the earth's atmosphere. This is due to the rising GHG concentrations in the atmosphere trapping increasing amount of the heat energy striking the earth from the sun plus heat emissions from different regions globally. These trapped heat energies were not able to be dissipated back into space and have led to a rise in global warming.

It is important to know that the earth's atmosphere has been acting like a giant greenhouse which has been capturing the sun's heat for many years. This has helped to create the right conditions and temperatures to support the emergence of different life forms, including humans, on earth. Without the earth's atmospheric greenhouse effect, the planet would become very cold. Scientists have estimated that the earth's temperature could fall by over 30 degrees Celsius, which would make many regions uninhabitable. On the other hand, rising GHG emissions starting after the Industrial Revolution have also led to serious global warming. Climate change and GHG emissions have led to the temperature of the earth going up faster now than at any other time before in history.

GHGs are major contributors to global warming and climate change. They are commonly defined as gases that have been emitted into the atmosphere which could absorb various infrared radiation and heat emissions. These GHGs are major contributors to the increased global warming. The key GHGs would normally include methane, CO_2, CFCs, etc.

Environment and sustainable development management have become important parts of climate change management by companies and cities globally. Environment and sustainability first became an integral part of the global climate change discussions at the World's first Earth Summit in Rio in 1992. There was no universally agreed definition on what environment and sustainability really entailed. There were many different views on what these would involve. The original definition of sustainable development, which is still the most often quoted definition, had come from the UN Brundtland Commission, which stated that 'Sustainable Development is development that meets the needs of the present without compromising the ability of future generations to meet their own needs' (UN Brundtland Commission, 1987).

Another good definition of environment and sustainable development is that of the World Commission on Environment and Development. It is expressed as 'A process of change in which the exploitation of resources, the direction of investments, the orientation of technological development and institutional change are all in harmony and enhance both current and future potential to meet human needs and aspirations.'

In practice, many leading international and state companies have been developing environmental and sustainable development strategies as an integral part of their corporate strategy. Global experience has shown that companies cannot just add environment and sustainable development to their list of corporate actions but they must integrate these into their core strategy. Many leading international companies have developed detailed environment and sustainable development strategies to support the sustained growth of their businesses. These strategies will normally aim to pursue simultaneously the three sustainable development pillars which include healthy environment, economic prosperity and social justice. Global experience has shown that these three sustainable development pillars should be pursued simultaneously to ensure the sustained business growth plus the well-being of current and future generations.

Climate changes and global warming overviews

Climate changes have led to large-scale, long-term changes in the earth's weather patterns and its average temperatures. Looking back, the earth's climate has been constantly changing over the past 4.55 billion years. Since the last ice age, which ended about 11,000 years ago, the earth's average temperature has been relatively stable, at about 14 degrees Celsius. However, in recent years, meteorologists have measured significant rises in global temperatures. These show that the global climate has been getting warmer and the average temperatures of the earth have been increasing steadily. They have also noticed that the rates of climate changes have been progressing faster with various human activities. These have contributed to accelerating the rates of climate changes and global warming worldwide. Climate change and global warming have resulted in many regions around the world suffering climate-induced extreme climate incidents, such as hurricanes, heavy rainfall, flooding and droughts. In addition, climate changes have also created severe problems for many plant and animal species globally. Some of these species have not been able to adapt fast enough to cope with the various serious climate-induced problems and are facing serious extinction risks.

Scientists have shown that climate change has been caused by human activities, which include rising global industrialisation, growing population, manufacturing and transport activities worldwide. Scientists have been studying the long-term relationships between atmospheric GHG levels, especially CO_2 concentrations, and the rise in global temperatures. Their results showed strong correlations between rising GHG levels, especially CO_2 levels, with global warming and industrial plus human activities. They have shown that global temperature rises commenced in the late 18th century in line with the start of the Industrial Revolution. The rising industrialisation in the 20th century and the 21st century has further contributed to rising GHG emissions, which have accelerated climate change and global warming.

GHG emissions have been shown to be the key cause for global warming. GHGs have been generated by industrial and agricultural activities plus human activities. The key GHGs include water vapour, CO_2, methane, nitrous oxide and CFCs. The concentrations of water vapour globally have shown few changes over history. In addition, water vapour usually lasts only a few days in the earth's atmosphere. On the other hand, CO_2 can persist for much longer in the earth's atmosphere and it can take hundreds of years for its concentrations to return to pre-industrial levels. Most man-made emissions of CO_2 have been generated by the burning and combustion of various fossil fuels, such as coal and oil fuels. In addition, the cutting down of forests for wood and fuel has reduced the earth's ability to re-absorb CO_2 by the trees. Other important GHGs such as methane and nitrous oxide have also been generated by various human and industrial activities (US EPA, 2019).

Scientists monitoring the global CO_2 levels have reported that since the start of the Industrial Revolution in 1750 CO_2 levels have risen by more than 30 per cent globally. The current concentrations of CO_2 in the earth's atmosphere have been found to be higher than at any previous times in history for the last 800,000 years. The concentrations of another important GHG, methane, have also risen significantly, by more than 140 per cent. Methane is emitted from coal bed methane mines or natural gas reservoirs. In addition, methane is also emitted by various agricultural animals, especially cows.

Extensive climate studies have shown that rising industrialisation and increased human activities globally have resulted in the build-up of man-made GHGs in the earth's atmosphere. These rising atmospheric GHG concentrations will then re-absorb the solar energy radiating back out to space from the earth's surface. The trapped heat energy and radiations would then be radiated back down to earth which would then heat up both the lower atmosphere and the surfaces of the earth. These would all contribute to the rising global warming and various serious climate change effects globally (Royal Society, 2014).

If there were no greenhouse effect and all the radiation was to be emitted back into space from earth, then the earth could become much colder than it is today, with our world's average temperature reducing by some 30 degrees Celsius. The colder environment would be hostile to many plant, insect, animal and human life forms that we know today. So it is very important that urgent joint climate change actions should be taken globally to protect the fine balances in the earth's atmosphere. We have to actively manage the GHG emissions and the subsequent negative climate change effects globally. These would need to be controlled and optimised so as to ensure the sustainable development of the planet and continued survival of various important life forms on earth.

Climate change and global warming major impacts

Studies have shown that climate change and global warming have resulted in major long-term negative impacts and disruptions around the world. Global climate studies have also shown that global warming and climate changes worldwide have been progressing more rapidly in recent years than in the past. The rising climate change effects have led to increased global warming and serious disruptions to global weather patterns. It is important to note that the global weather and climate systems are all inter-connected and finely balanced. As a result of global warming and climate changes, extreme weather events and disasters have occurred with rising frequencies, around the world. These have included hurricanes, flooding, drought, extreme rainfalls and so on (NASA, 2014).

The key impacts of climate changes experienced globally have included global warming, melting glaciers, rising sea levels, flooding, worsening droughts, hurricanes, supercell storms, increased tornados, extreme temperatures, heavy rainfalls, extreme weathers, larger seasonal changes, retreating glaciers, Arctic ice declines, ice sheet shrinkages, etc. The various key climatic impacts will be described in detail below, with relevant international examples.

Studies of the recent changes in the earth's temperature profiles have shown that the average temperatures of the earth's surface have risen by around 0.89 degrees Celsius from 1901 to 2012. When scientists compared these global warming patterns throughout the history of earth, they have found that the rates of temperature rise since the start of the Industrial Revolution have been very high. A good example is that since the last ice age, which ended some 11,700 years ago, the earth's averaged temperature has been relatively stable at about 14 degrees Celsius. However in recent years, after the start of the Industrial Revolution, measurements have shown that the world's climate has been getting warmer and the average temperatures of the earth have been increasing steadily. They have also noticed that the rising climate change effects have been driven by industrialisation and increased human activities. These have speeded up the rates of GHG emissions and the rates of global warming.

One of the most serious climate change impacts is global warming. This is driven by the higher re-absorption and trapping of heat energies on earth by GHG emissions. This has led to rises in the global average temperatures and the heating up of oceans around the world. These have then led to more water being evaporated from the different oceans worldwide into the atmosphere, causing more clouds to form. These have then led to the generation of more climate-induced extreme weather incidents, including high energy intensive storms, hurricanes and typhoons which have been hitting different regions around the world recently. A good example is that global warming has caused air to warm up over warmer seas. These caused the warmer air to rise rapidly and quickly develop into major storms and hurricanes. In general, the warmer the sea, the warmer the surrounding air will be. These will then make the hurricanes and storms worse. Weather statistics in the USA have shown that the number and intensity of hurricanes recorded have been rising recently. Between 1975 and 1989 there were 171 severe hurricanes recorded in the USA. The frequency of severe hurricanes has increased significantly and nearly doubled between 1989 and 2005, with 269 severe hurricanes recorded in the USA. Hurricane Katrina, which hit New Orleans in 2016, was one of the most severe hurricanes in US history. Katrina caused widespread disruptions and damages. Additional examples of recent hurricanes include the tropical Hurricane Harvey, which hit Florida in August 2017, plus Hurricane Irma, which hit the Caribbean and Florida in September 2017. All these tropical hurricanes have caused widespread damages with serious negative impacts (US National Hurricane Centre, 2018).

The rising amount of water vapour in the earth's atmosphere, resulting from global warming and climate changes, has also seriously changed the worldwide rainfall patterns. Scientists have found that significant changes in rainfall and precipitation have occurred in recent years. Their studies have found that the levels of rainfall have increased significantly in the mid-latitudes of the northern hemisphere since the beginning of the 20th century. They have also found more violent climate-induced heavy downpours have been taking place instead of steady normal showers around the world. Torrential downpours and more powerful storms have occurred in different parts of the world, especially in Asia and the Indian subcontinent. There has also been strong evidence that extreme heavy rainfall events have become more intensive and frequent, especially in America and Asia.

A good example of climate-induced rainfall was the heavy rainfall in the Indian subcontinent during August 2017, which led to serious flooding in various Indian subcontinental regions. The widespread flooding caused serious damage and disruption which affected over 40 million people in the Indian subcontinent.

Global warming and the changes in temperatures around the world have also changed the patterns of wind and air flows globally, leading to weather changes in different regions. A good example is Africa, where these have led to prolonged heatwaves becoming more common, which in turn has led to serious droughts in various African countries. Another good climate-induced example is that the Sahara Desert has been expanding more quickly in recent years due to the heatwaves and droughts caused by climate changes. A serious drought in Ethiopia has become one of the worst droughts globally, affecting millions of people and generating millions of climate refugees in Africa.

At the other extreme, climate changes have resulted in record cold temperatures and winter storms in many regions. Global warming has led to shifts in cold upper air currents as well as hot dry ones. Increasingly, cities that were once in the temperate zone, with regular rainfall are becoming hotter and drier. The string of record high temperatures in recent years and the higher occurrence of droughts in the past decade have become the norm, due to climate change.

Climate changes have also resulted in significant changes in seasons across the world in different countries. A good example is that the spring seasons in UK have been starting earlier and the autumn seasons have been starting later. There have also been changes in seasons in various regions globally. A good example is that the UK's summer rainfall has decreased on average whilst winter rainfall has increased. In recent years, the lower summer rainfalls in different regions in Europe and America have led to hose-pipe bans being imposed by various governments to conserve the limited available water resources. The reduced water availabilities have in turn resulted from the lower than normal rainfalls induced by climate change.

These climate-induced seasonal changes have also led to significant changes in the behaviour of different bird and animal species around the world. Two good examples are that butterflies have been appearing earlier in the year and various bird species have been shifting their migration patterns globally.

Climate change and global warming have led to rises in sea levels globally. Ocean studies have shown that since 1900, the sea levels have risen by about 10 cm around the UK and about 19 cm globally, on average.

The rates of sea-level rises have also increased in recent decades. The average sea level rises around the world have been about 8 inches (20 cm) in the past 100 years. Ocean scientists are expecting global sea levels to rise further and more rapidly in the next 100 years, due to climate change. Coastal cities around the world have already seen an increased number of flooding events. Looking ahead to 2050, many coastal cities would require extra seawall defences and flooding protection to protect against the sea level rises and to minimise serious flooding damage. Scientists have forecasted that sea levels globally would rise by 1 to 4 feet (30 to 100 cm) globally. These would lead to serious flooding in many small Pacific island states such as Vanuatu as well as important coastal cities, such as Bangkok and Boston. Serious flooding incidents around the world, which used to occur once every 100 years, have been forecasted to occur 5 to 10 times more frequently, say between every 10 to 20 years (NASA, 2019).

Global warming and the warmer atmospheric temperatures have also led to the melting of glaciers, mountain snow packs, the Polar ice cap, and the great Antarctica ice shield. A serious development is that the Inuits in the Arctic regions have noticed that the Arctic ice sheets have been melting more rapidly in recent summer months and freezing less in the winter months recently. Polar studies have shown that since the 1970s, the Arctic sea-ice coverage has been declining by about 4 per cent, or 0.6 million square kilometres per decade. Converting this to equivalent land areas, the accelerated Arctic ice sheet melting would be equivalent to the losing of a large island with an area about the size of Madagascar.

The Antarctica, Arctic and Greenland ice sheets have all been affected by climate change and global warming. They have been storing the majority of the world's fresh water reserves. The melting of these ice sheets, as a result of global warming, has contributed directly to rising sea levels globally.

Scientists have also shown that the ice caps at both the Arctic North Pole and the Antarctic South Pole have been melting at higher rates in recent years as a result of global warming. These have in turn caused sea levels to rise globally. Scientists have forecasted that if the Antarctic ice shelf and the Greenland ice sheets were both to melt due to global warming, then sea levels globally could rise by as much as 6 metres or 20 feet. This would then lead to large parts of the US Gulf coast, Florida and New Orleans plus Houston all being flooded and suffering serious flooding damage, which could affect millions of people and their homes. (Glick, 2019).

Climate changes and global warming have also led to ice glaciers retreating all over the world. Glacier studies globally have shown that the ice glaciers in different countries across the world, including the Alps, Rockies, Andes, Himalayas, Africa and Alaska, have all been melting at increasingly alarming rates. The rates of shrinkage of various major ice glaciers globally have increased alarmingly in recent decades. Some scientists have projected that climate change could lead to most of the world's glaciers disappearing within the next 100 years. Looking ahead to 2100, most of the ice glaciers around the world might have melted due to global warming. This would lead to serious negative ecological and environmental impacts on the glacier regions globally (NSIDC, 2019).

Scientists have also been developing advanced climate computer models to show and forecast how climate changes and global warming would affect the entire ecological systems and global weather systems. These should improve forecasts of various potential extreme climate incidents globally, including sea level rises, hurricanes, typhoons, extreme temperatures, heavy rainfalls, flooding, droughts, ice cap melting etc. These should help to minimise serious disruptions and impacts of these serious climate-induced incidents. Good examples of serious climate-induced extreme events include the severe droughts in Africa, which have caused many human deaths, the changing ocean pH and acidity plus rising sea temperatures which have resulted in major shifts of fish movements and fish species in the oceans around the world and so on. These climate models have helped scientists in their studies on global warming and enabled them to predict the negative outcomes of climate change globally. More details on these new global climate models will be provided in Chapter 2 (UN IPCC, 2013).

Climate change and global warming major causes

Various scientific studies have shown that there are many factors which have contributed to climate changes globally. In general, any actions or substances that could influence and affect the amount of energy being absorbed from the sun or the amount of radiation being emitted by the earth could lead to long and short-term changes in global warming which impact the earth's climate systems. These could then lead to a serious rise in global temperatures. Studies have shown that global warming has been caused by the rising emissions generated by industrial and human activities. These have increased the amount of GHGs in the earth's upper atmosphere and the amounts of black carbon and tiny particulate pollutants in the earth's lower atmosphere (European Commission, 2019).

GHG emissions have been generated by the burning of various fossil fuels, including coal, oil and gas, in industrial and human activities globally. Tiny carbon particles and black carbon soot have been produced by the incomplete combustion of fossil fuels, especially from coal and diesel. These pollutants have contributed significantly to global warming because they trap the sun's energy in the earth's atmosphere. GHGs have been acting like the reflective glass in a greenhouse which trap heat radiations from earth. The rising amounts of tiny 'black carbon' particles or soot, generated by fossil fuel combustion, in the earth's atmosphere have contributed significantly to global warming. The resultant layer of fine black particles in the lower atmosphere of the earth has been absorbing heat and radiation like a black blanket covering earth. More specific details on GHGs and black carbon will be discussed later in this chapter.

Studies have shown that global warming started in late 18th century, when coal was first brought into common industrial and residential uses. Since then coal has been burned in many power stations globally. Coal has also been burned in many residential houses for heating and cooking, as well as being used to power trains and ships. The widespread industrial and residential combustion of coal has led to massive emissions of GHG and particulates into the earth's atmosphere. These have contributed to the onset of serious global warming in the late 18th century which has since continued into the 19th, 20th and 21st centuries (US EPA, 2016).

Studies have shown that global warming trends have accelerated in the late 19th to 21st centuries due to the increased global use of various petroleum fossil fuels, including gasoline, diesel, kerosene, bunker and so on. These fossil fuels have been applied in a range of industrial, power generation, manufacturing and transport settings globally. These have led to increased GHGs and particulates emissions from power stations, factories, cars and trucks, etc. The accumulation of GHG and pollutants emissions in the earth's upper and lower atmospheric layers have led to global warming. Additional GHGs, especially methane, have also been generated by other traditional human and animal activities such as farming and agriculture.

Studies by international energy agencies have shown that the large majority of global energy consumption, estimated to be some 80 per cent, has been derived from fossil fuels. These include coal, oil, natural gas, shale gas, gasoline, kerosene, diesel etc. It was reported that in 2013 that fossil fuel combustion and uses globally have given rise to 32.3 Gt per annum of global energy-related CO_2 emissions. In 2014, similarly high levels of greenhouse emissions were also recorded. However it was also the first time in 40 years of CO_2 global measurements that there was a levelling out of GHG emissions which was not linked to an economic downturn globally. The levelling out of global CO_2 emissions largely resulted from new controls and clean energy policies introduced by various governments following higher global awareness of climate change. These helped to arrest the rising CO_2 emission trends for the first time in the last 40 years. Unfortunately this was short-lived, since global CO_2 emissions went up again in the 2016–2018 period. Scientists have warned that one of the biggest CO_2 surges in more than six decades of measurements might occur in 2019–2020 (Gabbatiss, 2019).

Climate change and global warming by GHGs

Scientists have provided important evidence that one of the key drivers for global warming has been the rising emissions of GHG and particulates globally. GHGs are responsible for the greenhouse effects in the earth's atmosphere which have led to global warming. The most well-known GHG is carbon dioxide, also known as CO_2. Other important GHGs include methane, carbon monoxide, sulphur dioxide, nitrous oxide and water vapour.

GHGs have not always been the bad players for the earth's climate. The earth actually needs some GHGs in the atmosphere in order for life on earth to exist. Trees and plants cannot survive without CO_2 as they will need it for photosynthesis. The plants will in turn provide food plus generate oxygen. The right amount of greenhouse gases will also help to keep earth warm enough for life to exist. Without GHG, the world could be some 30–33 degrees Celsius colder than it is now and many life forms would not be able to survive (CIFOR, 2019).

A good climate example is that the average temperature for November in the UK is around 6 degrees Celsius. If there were no GHGs, then the temperature would drop to a freezing –27 degrees Celsius instead. This is because the GHGs would form a protective layer in the earth's atmosphere which would stop all the sun's radiation and warmth being lost back into space. The problem now is that emissions from various human and industrial activities have increased greenhouse concentrations in the atmosphere to much higher levels than they should be naturally, which could lead to increased global warming plus more extreme weather incidents.

One of the most important GHGs on earth is water vapour. As the earth's atmosphere warms up due to global warming, then the amount of water vapour in the atmosphere would also increase. Nature has been regulating the fine balance of water vapour in the atmosphere with clouds and rain. Recently the rising global warming and climate change have led to more extreme rainfalls and extreme weather incidents globally.

CO_2 and methane (CH4) are both important GHGs. Both of these GHGs have strong 'forcing' effects which could increase the effects of global warming. Their rising concentrations have been caused by industrial and human activities. There have also been significant potential secondary effects. A good example is the significant reduction in natural carbon storages due to the rising human logging activities which have caused widespread deforestation globally. In addition, global warming and melting of the ice sheets might lead to the release of methane hydrates which have been trapped in the permafrost layers for thousands of years (World Ocean Review, 2010).

CO_2 is the most common GHG on earth and has constituted almost over half of the total GHGs in the atmosphere. The amount of CO_2 in the atmosphere has increased dramatically over recent years. Scientific measurements have shown that CO_2 concentrations globally have risen by about 38 per cent from the start of the Industrial Revolution to the 21st century. The continued burning of fossil fuels and rising industrial activities have contributed to the amounts of CO_2 globally continuing to rise. These have contributed to absorptions of more radiation and heat in the earth's atmosphere. This has led to fewer emissions of the earth's radiations, which has resulted in more heat being trapped in the earth's atmosphere, leading to further rises in global warming.

CO_2 has been used as a marker by the United States Environmental Protection Agency (USEPA) because of its ubiquity and has been assigned a

Global Warming Potential (GWP) of 1. CO_2 in the atmosphere could also last a long time. Studies have shown that CO_2 could last 100 years or more in the earth's atmosphere. This means that CO_2 has a very long time to build up and influence the earth's climate. Scientific carbon dating measurements have shown that some of the CO_2 currently in the earth's atmosphere was emitted before World War I. Human activities, especially the cutting down of forests, have increased the imbalances between the rising CO_2 emissions and the earth's natural capacity to re-absorb it (US EPA, 2018).

Methane or CH4 is a strong GHG which could seriously affect global warming. In contrast to CO_2, methane normally will stay in the atmosphere for more than about a decade. US EPA has shown that methane can have 28 to 36 times greater impacts on global warming than CO_2 in earth's upper atmosphere. This means that 1 tonne of methane is equivalent to 28–36 tons of CO_2 in terms of its global warming impacts. So in comparison to CO_2, CH4 has been assigned a GWP of 28–36 whilst CO_2 has a GWP of 1.

Methane is being produced in many fossil fuel combustion processes and industrial process. In addition, it is produced by anaerobic decomposition from various farming and agricultural activities, such as from flooded rice paddies, pig and cow stomachs, and pig manure ponds. Methane will normally break down in approximately 10 years in the earth's atmosphere. Methane is also a precursor of ozone which is another important GHG. Looking ahead, global warming and increased melting of the polar ice sheets could cause the potentially very large releases of the methane hydrates reserves which have been trapped in the ice permafrost layers for the last thousands of years.

Nitrous oxide (NO/N2O or simply NOx), is also called laughing gas and is an important GHG. It is normally produced as a by-product from fertilizer production and agricultural applications. It can also be generated by industrial processes, combustion processes and vehicle exhausts. Nitrous oxide can last a very long time in the earth's atmosphere. In comparison to CO_2, nitrous oxide has a very high GWP of 265–298. This means that 1 ton of nitrous oxide could cause as much global warming as 265–298 tons of CO_2.

Fluorinated gases have been produced as replacements for the older ozone depleting refrigerants. Fluorinated gases are important GHGs which are long lasting, with big global warming impacts. The fluorinated gases normally have no natural sources and have been produced entirely from man-made industrial sources. In comparison to CO_2, fluorinated gases have been assigned GWPs ranging from 1,800 to 8,000 and some variants even reached 10,000. This means that 1 ton of fluorinated gas could have as much impact on global warming as 1,800 to 10,000 tons of CO_2, which is very high.

Sulphur hexafluoride or SF6 is used for specialized medical procedures and is an important GHG. It is normally present in di-electric materials, especially dielectric liquids. These have been used as insulators in high voltage applications such as transformers and grid switching gear. SF6 could last thousands of years in the earth's upper atmosphere. They also have high global warming potentials and have been assigned a very high GWP of 22,800. This means that 1 ton of SF6 gas will have as much impact on global warming as 22,800 tons of CO_2.

All these GHG emissions have contributed significantly to global warming and climate changes globally. In particular, energy-related CO_2 emissions have been responsible for the majority of global GHG emissions and global warming. The various approaches to GHG emission management and reductions will be discussed in later chapters.

Climate change impacts of CO_2

Climate change and global warming have been described as two of the biggest problems faced by humankind. CO_2 has been shown to be one of the primary drivers of global warming and climate change globally. CO_2 has been present in the atmosphere since the formation of the earth about five billion years ago. At that time the earth's atmosphere was mainly composed of nitrogen, CO_2 and water vapour. These are similar to the gas compositions emerging from volcanic eruptions today. As the earth cooled further some of the water vapour condensed out to form oceans. These have helped to dissolve a portion of the CO_2 but the rest of the CO_2 is still present in the atmosphere in large amounts. About 2.5 billion years ago, microbes and plants developed the ability to photosynthesis, creating glucose and oxygen from CO_2 and water in the presence of light from the sun. As life developed, CO_2 was consumed and around 20 million years ago its concentration was down to below 300 molecules in every one million molecules of air (or 300 parts per million (ppm).

When human beings appeared on earth about 200,000 years ago, the atmospheric CO_2 concentrations continued to stay relatively stable for many years around the 280 to 300 ppm level. Climate studies showed that until the start of the Industrial Revolution, the earth's CO_2 concentrations have generally not exceeded the 280ppm value. During the last ice age, declines in the global temperature of perhaps 5 degrees Celsius have been accompanied by reductions in CO_2 concentrations to less than 200ppm. With the start of the Industrial Revolution, the earth's CO_2 concentrations have started to rise. These CO_2 emissions have been caused by the various industrial and human activities, particularly with the combustion of fossil fuels. The GHG emissions were particularly high from the combustion of coal and fossil fuel.

It is important to recognise that CO_2 plays a very important role in the climate of the earth. It is one of the important atmospheric 'greenhouse' gases (GHGs) which has helped to keep the earth at the right temperature for life. Scientists have estimated that the right CO_2 concentrations have helped to keep the earth about 33 degrees warmer. Without CO_2, the earth would be much colder at a temperature of −18C which would be too cold to support many of the current life forms on earth. CO_2 is able to achieve this by being fairly transparent to the sun's rays. This allows the sun's radiation to pass through the atmosphere to warm up the surface of the earth. CO_2 also absorbs the radiant heat emitted by the Earth's surface, which contributes to global temperature rises and global warming.

In the earth's finely balanced climatic systems, one of the most important and effective GHGs has been water vapour. It is estimated water vapour has been responsible for about two-thirds of the total warming effects. In comparison CO_2 had accounted for about one quarter of the total warming effects. Other GHGs, including methane and fluorinated gases etc., have contributed to the remainder of the global warming effects.

GHGs have big impacts on global warming because their molecular structures have made them very effective at absorbing heat radiations. In comparison, the other major atmospheric gases, such as nitrogen and oxygen, have molecular structures which make them essentially transparent to heat.

The important GHG effects mean that as the atmospheric concentrations of GHGs rise, then the surface temperature of the earth rises further more. Climate scientists have found that the overall increases in the global temperature of about 1 degree Celsius over the past 150 years have been almost entirely due to industrial and human activities which have increased the amounts of atmospheric GHGs. Most significantly, the concentrations of CO_2 in the earth's atmosphere had been rising exponentially, at rates of about 0.17 per cent per year, since the start of the Industrial Revolution. These have been mainly caused by the combustion of fossil fuels, especially coal and oil. In addition, large-scale deforestation activities in various countries have also helped to reduce the earth's natural capacity to reabsorb CO_2 via photosynthesis. In 2015, global CO_2 concentrations passed 400ppm. This was more than 40 per cent higher than the pre-industrialisation CO_2 concentration of 280ppm. These high CO_2 levels have not existed on earth for several million years. A good evidence for climate change is that the air sampling station at the Mauna Loa observatory in Hawaii has recorded atmospheric CO_2 concentrations rising past 400ppm.

Whilst the basic science of how GHGs could contribute to global warming has been quite well understood and researched, there are several important climate change complications. The earth's complex climate systems have been found to respond in various complicated ways which could both enhance and ameliorate the effects of these GHG gases. A good example is that a warmer atmosphere could also lead it to hold more water vapour before these are condensed out in clouds or rain. As water vapour is a strong GHG, this would also contribute to further temperature rises and global warming. In addition, eruptions of volcanoes in various countries could lead to emission of major amounts of GHGs plus small soot particles into the upper part of the atmosphere. These would then lead to further global warming and climatic changes around the world. Another important climate change impact has been that as the atmospheric CO_2 concentrations rose then these have led to increased acidification of oceans around the world via more CO_2 dissolving in seawater. These have in turn led to sea water pH changes which have serious climate-induced implications on sea life and coral around the world. These climate change implications and outlooks will be discussed in detail in the next chapter (NOAA, 2018).

Climate change impacts of fossil fuels and black carbon

Various climate and pollution studies globally have shown that the rising uses of fossil fuels, especially coal and petroleum, have been the most important sources of GHGs and black carbon emissions to the earth's atmosphere. Fossil fuels have been used in many industrial and residential activities, including power generation, industrial processes, chemicals manufacturing, transportation and buildings etc. The combustion of fossil fuels has resulted in the generation of significant amounts of GHGs including CO_2, nitrous oxide plus black carbon, soot and particulates emissions to the earth's upper and lower atmospheres.

It is also important to recognise that agriculture and farming activities have represented the second largest sources of GHG generations globally. These would include waste gas emitted by farm animals, such as cows and pigs etc. Farming activities, such as intensive chemical fertiliser food production, field burning and flooded paddy rice production, have also been generating high GHG emissions. In addition, widespread deforestation and torching of woodlands in various countries, such as Indonesia and Malaysia etc., have also led to high GHG emissions plus big reductions in the earth's natural carbon reabsorption capacities. Many climate studies have shown that the agricultural sector has been the largest contributor of particulate emissions and pollution in the USA and other agricultural countries.

Black carbon or BC contains tiny particles of carbon generated by the incomplete combustion of fossil fuels, biofuels and biomass. These particles are extremely small, ranging from 10 µm (micrometers, PM10) to less than 2.5 µm (PM2.5). These small particulates are normally harmful to human lungs and respiration systems. In particular the smaller particulates PM2.5, which is one thirtieth the width of a human hair, have been considered to be most harmful. They are small enough to pass through the walls of the human lung into the human bloodstream leading to serious human health problems.

Black carbon emitted from the plume of smoke from a chimney or a fire would normally fell out of the lower atmosphere in days. However whilst these particulates are being suspended in the air, they could absorb the sun's heat rays millions of times more effectively than CO_2. If the black carbon particulates were carried by wind and deposited over snow, glaciers or ice caps then these could cause even more damages. When the black carbon particulates fell out of atmosphere onto the white, reflective surfaces of ice or glacier, it would contribute directly to higher heat absorption and faster melting. Overall, black carbon BC particulates are considered to be the second biggest contributor to global warming just after CO_2 (Nagoya University, 2018).

The electricity power generation sector globally has been a major fossil fuel consumer and has generated significant amounts of GHG and particulate emissions. In 2015, 68.5 per cent of the world's electricity production was generated from fossil fuels power combustion plants. These included coal, oil and gas combustion. In some Middle East countries, they have been burning oil and petroleum in their electricity generation plants. In comparison, hydro-electric plants only provided 16 per cent of clean electricity globally. Nuclear plants generated slightly over 10 per cent of electricity globally. Renewable energies, including geothermal, solar and wind have generated only 4.9 per cent of the world's electricity in 2015. Biofuels and waste gases generation made up the remaining 2.2 per cent.

Between 1974 and 2015, the world's gross electricity productions have increased significantly from 6,287 TWh to 24,345 TWh. These significant power generation growths have represented an average annual growth rate of 3.4 per cent globally. In 2015, electricity production was 1.7per cent higher than in 2014. These represented the sixth straight year of rising electricity growths, after the 2008–2009 economic crisis in OECD countries which caused a temporary decline in global electricity productions.

Since 2000, the higher economic growths of various non-OECD emerging economies have increased the share of non-OECD countries in total world electricity productions. From 1974 to 2000, non-OECD countries electricity productions have increased at an average annual rate of 4.6 per cent, whereas OECD countries have only increased at 3.0 per cent. These global trends changed significantly from 2000 to 2015. The average annual growth rates fell to only 0.9 per cent in OECD countries whilst they grew by 5.9 per cent in non-OECD countries. Consequently, in 2011, non-OECD electricity productions exceeded OECD electricity production for the first time. The non-OECD power growths have been continuing. In 2015 non-OECD countries accounted for over 55 per cent of the world electricity generation. This was almost double its share of 28.1 per cent in 1974. The future outlook of the world's rising electricity generation requirements plus their implications for climate change and global warming will be described in more detail in the author's planned new book on renewable energy growths.

2 Climate change global outlooks and implications

吃得苦中苦, 方为人上人
chī dé kǔ zhōng kǔ, fāng wéi rén shàng rén
A person who can endure hardships, will become a better person.
No pain, no gain.

Executive overview

Industrialisation and fossil fuel consumptions have contributed to significant rises in GHG emissions and climate change problems globally. Looking ahead to 2100, the various climate change impacts are likely to result in higher temperature rises and more climate-induced extreme weather incidents, including flooding, heavier rainfall, hurricanes, droughts, etc. These are likely to cause major disruptions to various cities globally with major financial costs. Leading scientists globally have all warned of a potentially very serious climate change tipping point coming up globally. After the tipping point, the damages caused by global warming and climate change would then become irreversible. Details of the various climate change impacts and outlooks will be discussed further in this chapter, with international examples.

Climate change and global warming outlooks

Climate changes and global warming have led to large-scale, long-term changes in the earth's weather patterns and more extreme weather incidents affecting various cities. The earth's climate systems have been constantly changing over the last 4.55 billion years. After the last ice age, which ended about 11,000 years ago on earth, the global averaged temperature has been relatively stable at about 14 degrees Celsius. However, in recent years, meteorologists have noticed some significant changes in the earth's temperature measurements. These showed that the average temperatures of the earth have been increasing steadily due to global warming. Scientists have also noticed that the rates of climate changes have still been rising with various industrial and human activities. These have been shown to have contributed directly to accelerating the rates of climate changes and global warming worldwide. The global climate changes have resulted in many regions around the world suffering severe weather conditions and climate-induced extreme weather incidents. In addition, climate changes have also created conditions which are difficult for many plants, insects and animal species to survive as they have not been able to adapt quickly enough to the fast-changing climate conditions.

Looking ahead, leading climate institutes globally have predicted further significant growths in global emissions due to rising industrialisation and population growth. Looking ahead to 2100, they forecasted that CO_2 emissions could further rise alarmingly by over 60 per cent when compared to the CO_2 emission levels in 2010. Looking ahead to 2050, the emissions from developed countries have been forecasted to reduce to about 15 per cent of global emissions, which is half of their 30 per cent share in 2010. On the other hand, the CO_2 emissions from developing countries have been forecasted to increase significantly by 2050 due to their higher economic growths and industrialisation rates (MIT, MIT Climate Action, 2017).

Experts have also predicted that CO_2 emissions from fossil fuels usages globally will continue to remain as the largest source of GHG emissions. Other GHG emissions and non-fossil energy sources of CO_2 will likely account for almost 33 per cent of total global GHG emissions by 2100. The emissions from the electricity generation and transportation sectors will likely account for over half, about 51 per cent, of the global CO_2 emissions. This would represent a slight decrease from the 56 per cent emission levels recorded in 2010.

Looking ahead to 2050, experts forecasted that fossil fuels would still account for about 60–75 per cent of primary energy usages around the world, despite the rapid growths in renewable energies and nuclear energy. Within the fossil fuel primary energy mix, there would be a strong shift away from coal and oil to cleaner fossil fuels, such as natural gas. The natural gas share in the primary energy mix in most countries globally would increase strongly. The rising natural gas applications have mainly been driven by the new cleaner energy policies introduced by different governments to promote clean fossil fuel usages and to reduce GHG emissions as part of their Paris Agreement commitments.

Looking ahead, the continued global warming and climate change would likely cause more frequent climate-induced severe weather events globally, including hurricanes, extreme precipitation, sea level rises and ocean acidification. Global mean surface temperatures have been forecasted to further increase by a range of 1.9 to 2.6 degrees Celsius by 2050. Looking ahead to 2100, the global mean surface temperatures have been forecasted to rise even further by a range of 3.1 to 5.2 degrees Celsius, if no joint global actions would be implemented to achieve the IPCC 1.5C target. By 2050, the global mean precipitation or rainfalls would likely increase by a range of 3.9 per cent to 5.3 per cent. Looking ahead to 2100, it would further rise to 7.1 to 11.4 per cent as a result of continuing climate change effects globally (MIT, MIT Climate Action, 2017).

Global warming has also resulted in serious melting of the polar ice cap and glaciers globally. These accelerated ice melting incidents are expecting to continue to increase globally. Looking ahead to 2100, experts have predicted many glaciers around the world might have melted and no longer exist.

Looking ahead, experts have predicted ocean levels to rise further globally. The thermal expansions of the oceans due to higher global temperatures plus the accelerated melting of ice sheets and glaciers would add a further 0.15 to 0.23 metres rises in sea levels by 2050 globally. Looking ahead to 2100, experts forecasted the global sea levels to rise further by 0.30 meters to 0.48 metres due to worsening climate changes and global warming. These large sea level rises globally would likely lead to more frequent flooding incidents globally, together with rising flooding risks for many coastal cities and islands globally.

Experts have warned that climate change and global warming are amongst the biggest threats to future sustainable human existence on earth. The late physicist Stephen Hawking and leading scientists globally have all warned of the risks of a very serious climate change tipping point materialising soon. They have warned that if the world were to go beyond the climate tipping point, then the negative effects of global warming and climate change that we have been observing would become more serious, with the damage becoming irreversible. According to climate scientists, we have been moving closer to the climate tipping point, with rising risk of the climate damage becoming permanent and irreversible. The UN's Intergovernmental Panel on Climate Change (IPCC) has also highlighted the potential serious risk of hitting the climate change tipping points as global warming worsens. There are really urgent needs for joint global actions to control climate change and reduce global warming before the climate tipping points are exceeded (BBC News, Stephen Hawking's Warnings, 2018).

Climate change carbon dioxide (CO_2) emissions outlooks

Studies have shown that carbon dioxide (CO_2) is one of the key GHGs contributing to global warming. It has been present in the earth's atmosphere since its formation millions of years ago. As life developed on earth, CO_2 in the atmosphere has been gradually consumed. Around 20 million years ago, the global CO_2 concentrations went down to below 300ppm or parts per million. These were some 100ppm lower than the current CO_2 concentration of over 400ppm. Atmospheric CO_2 concentrations have been stable for the last millions of years. The global CO_2 concentrations did not rise significantly until the start of the Industrial Revolution in the late 18th century. The Industrial Revolution led to higher emissions of GHGs and pollutants. These have been generated by the combustion of various fossil fuels, especially coal and oil, due to rising industrial plus various human and animal activities.

Scientists have shown that the concentration of CO_2 has been rising exponentially, at a rate of about 0.17 per cent per year since the Industrial Revolution. This has been mainly due to the combustion of fossil fuels plus various industrial and human activities. In addition large-scale tropical deforestation activities have also reduced the earth's carbon re-adsorption capacities via photosynthesis. In 2015, the global CO_2 concentrations rose passed the 400ppm levels. The current CO_2 concentrations have risen some 30 per cent to 40 per cent higher than the pre-industrial concentrations of 280–300 ppm. The latest WMO State of the Climate Report has reported that global CO_2 level has risen from 357 parts per million (ppm) in 1993 to 405.5ppm in 2019. Looking ahead, the global CO_2 levels are expected to increase further, resulting in worsening global warming (WMO, 2019).

Most man-made emissions of CO_2 have resulted mainly from the burning of fossil fuels, especially coal and oil, plus various industrial plus human activities. Other GHGs such as methane and nitrous oxides have also been released through industrial and agricultural activities. However, the concentrations of other GHGs are small when compared with the rising CO_2 levels globally. Since the start of the Industrial Revolution in 1750, CO_2 levels globally have risen by more than 40 per cent and the methane levels have risen by more than 140 per cent. The concentrations of CO_2 in the atmosphere are now higher than at any time historically for the last 800,000 years.

In 2017, the energy-related CO_2 emissions grew by 1.4 per cent globally. The global CO_2 emissions reached a historic high of 32.5 gigatonnes per year (Gt/y). Analysis of the CO_2 emissions records over recent years has shown that there has been an alarming resumption of CO_2 emission growths globally. CO_2 emissions have risen again recently after three years of global CO_2 emissions remaining flat. The increases in CO_2 emissions were not even globally but have varied greatly between countries. Most major developed economies have seen a rise in their CO_2 emissions. In contrast, some leading countries have experienced encouraging declines in their CO_2 emissions. These included the United States, United Kingdom, Mexico and Japan, which have been implementing new cleaner energy and climate change policies (IEA, 2019).

IEA has just reported, in their *Global Energy and CO₂ Status Report* of 2018, that the global energy-related CO_2 emissions have risen by 1.7 per cent to a record high global CO_2 emission of 33.1 Gt CO_2 in 2018. The emissions from all fossil fuels, especially coal and oil, have increased globally. These were driven by higher energy consumptions globally, which had increased at nearly twice the average rate of growths. Higher electricity demands were responsible for over half of the energy growth globally. The global fossil power generation sector has accounted for nearly two-thirds of the GHG emissions growths. Coal used in power generation alone generated over 10 Gt CO_2 emissions, mostly in Asia, China, India, and the United States. These contributed to some 85 per cent of the net increase in GHG emissions. There were encouraging emissions declines for Germany, Japan, Mexico, France and the United Kingdom with their clean energy policies and drives (IEA, 2019).

The increasing global CO_2 emissions have led to further global temperature rises and worsening climate impacts. A serious impact would be that as oceans warm up, they would also be expanding, leading to rising sea levels globally. These sea level rises have also been exacerbated by the melting of the polar ice sheets and glaciers globally. These would in turn lead to more flooding of cities and coastal regions globally. The warmer atmosphere would also hold more water vapour. This would lead to rising occurrences of extreme weather incidents, including heavier rainfalls and storms in different countries. At the other extreme, the changes in weather patterns globally have also resulted in intensifying droughts in some regions, such as Africa.

The rising CO_2 atmospheric concentrations have also led to increased CO_2 dissolved concentrations in various oceans globally. These have then led to rising acidity of oceans around the world. Ocean acidification is having serious impacts on ocean and fish life. Looking ahead to 2100, it is expected that the average pH of oceans globally could drop from 8.13 in 2010 to about 7.82 by 2100. These reduced ocean pHs would make the oceans more acidic, which would have serious impacts on the various fish and sealife ecosystems around the world.

Looking ahead, global CO_2 emissions could increase alarmingly by over 30 per cent by 2035. This is based on the scenario that the major nations globally do little to change their existing energy policies and do not take strong climate change actions to control global warming. This could then result in global energy consumptions increasing by close to 50 per cent by 2035. Most of the new incremental energy consumptions would be in the major emerging economies. These would include China, India and other key developing countries as their economic growths continue with more industrialisation, power generation and more cars.

There is a significant degree of uncertainty surrounding these long-term projections of energy-related CO_2 emissions. These could be heavily influenced by new energy and climate change policies in various key countries globally. These in turn could shift significantly depending on the future political developments in these countries. Changes in governments and election results could seriously affect governments' climate change policies and their commitments to their Paris Agreement obligations.

Looking ahead, some climate scientists have forecasted an optimistic climate change scenario based on new cleaner energy policies being introduced by many countries. Many countries have recognised that they cannot continue to pursue economic growths in the absence of sound climate and environmental policies. These should lead to reduced GHG emissions and lower global warming. A good example is that China has announced major investments on clean renewable energies applications in their 13th Five Year Plan which should help to lower their GHG emissions and reduce pollution in the coming decades.

At the other extreme, some climate scientists have predicted a more pessimistic scenario involving further significant growths in global CO_2 emissions. Looking ahead, they have forecasted that CO_2 emissions could rise to over 60 Gt per annum of CO_2-equivalent emissions by 2050 and then further rising to 78 Gt per annum by 2100. These would mean that the CO_2 emissions in 2100 could be a higher 63 per cent increase as compared to the CO_2 emission levels in 2010.

Looking ahead, experts have also forecasted that, despite the rising application of clean renewable energy, fossil fuels would continue to be the main dominant energy sources in the world energy mix for the foreseeable future. They forecasted that CO_2 emissions from fossil fuels usages globally would continue to remain as the largest source of GHG emissions globally. These future rising CO_2 emissions would likely lead to worsening global warming globally. Looking ahead to 2050, global mean surface temperatures could further increase by some 1.9 to 2.6 degrees Celsius. Looking ahead to 2100, the global mean surface temperatures could rise further by another 3.1 to 5.2 degrees Celsius. The rising CO_2 emissions and global warmings would also lead to more extreme weather incidents, including hurricanes and heavier rainfall globally. By 2050, global mean precipitation could increase by a range of 3.9 per cent to 5.3 per cent. Looking ahead to 2100, these could further rise to 7.1 to 11.4 per cent. Hopefully the new clean energy policies and climate measures that are being introduced by different countries globally could help to reduce fossil energy consumptions plus reduce GHG emissions.

In the future, there could be a range of possible climate change scenarios for the world. Which one will materialise in the end will depend on different key drivers, including fossil fuel consumptions, GHG emissions, industrial activities and renewable energy applications. The severity of the long-term climate disruptions would also be dependent on the future new energy policy and emissions controls that various governments would be introducing globally. As climate and weather are globally connected, the various climate change impacts would be felt everywhere globally.

The rising global warming and worsening climate change would likely lead to more severe weather incidents with potentially disastrous implications globally. These climate-induced extreme weather incidents could include extreme temperatures, heavy rainfall, hurricanes, ice melting, flooding and drought. The potential outlooks and implications of these climate-induced severe weather incidents will be described in more details below in this chapter, together with international examples.

Extreme weather hurricanes climate change implications

Climate change and global warming have generated many negative impacts to the earth's weather systems globally. One of the most serious climate-induced severe weather incidents has been the increased occurrence of extreme hurricanes in different regions of the world. These have caused severe damage and disruption to communities and cities.

Climate change and global warming have caused significant changes to the entire global weather system. These have resulted in significant shifts in the cold upper air currents as well as the hot dry currents on the earth's atmosphere. These have then contributed to more severe hurricanes developing in recent years. A recent example of a strong climate- induced hurricane was the tropical Hurricane Harvey which hit the US, Florida and Caribbean regions causing severe damage. Hurricanes are very complex, naturally occurring, extreme weather incidents which are extremely difficult to predict with any degree of certainty as these require complex climate modelling and extensive historical data.

Hurricane scientists have been investigating the precise role of climate change on the formation of severe hurricanes and their serious impacts. There is an important physical law, the Clausius-Clapeyron equation, which said that hotter atmosphere would hold more moisture. In general, every extra degree Celsius in global warming would lead to the earth's atmosphere holding an extra 7 per cent more water. Researchers have also reported that climate change and global warming have caused significant changes in atmospheric circulation in the mid-latitudes of the earth's atmosphere. The warming in the Arctic could also lead to the earth's weather systems moving less around the world and staying longer in some locations. These combined negative climate change impacts could then significantly enhance the formation of stronger hurricane with heavier rainfalls (UN IPCC, Fourth Assessment Report on Climate Change, 2007).

An example of a major region adversely affected by severe hurricanes was the Houston, Texas and Gulf of Mexico area in the US. Texas has been affected by many slow-moving hurricane storms. Recent climate-induced severe hurricanes affecting Texas have included the major serious tropical storms Claudette in 1979 and Allison in 2001, plus the recent severe tropical Hurricane Harvey. These have led to exceptionally strong winds and heavy rainfall in the Texas Gulf regions as they settled over the state for longer periods. Climate researchers have also shown that the intensity of the rainfall in the Houston and Gulf of Mexico areas have been strongly linked to climate changes. Ocean temperature measurements have shown that the waters of the Gulf of Mexico have risen by about 1.5 degrees Celsius during 1980–2010. These have contributed to the formation of stronger hurricanes and storms around the southern US and Gulf of Mexico regions.

To better understand the linkages of climate change with the formation of extreme hurricanes and storms, scientists have been using high-resolution computer models to study these. Climate scientists have been undertaking more computer simulations with actual hurricane and storm data inputs so as to better simulate and predict the formation of hurricanes. In addition, they have been undertaking more weather tracking of hurricanes, with professional hurricane chase teams, around the various US regions. Hopefully this new hurricane research will lead to better prediction of hurricane formation and improved hurricane warnings for communities. These hurricane incidents have also highlighted the importance of joint international climate change actions to reduce global warming so as to minimise the occurrence of climate-induced extreme weather incidents globally.

With the impacts of climate change worsening, some environmental lawyers have started to question if extreme weather events, like tropical Hurricane Harvey, should still be referred to as 'Acts of God' or 'Natural Disasters'. These lawyers have been arguing that these incidents have been created by greenhouse emissions from fossil fuels due to planned industrial and human activities. Some environmental lawyers have advocated that legal actions should be taken against countries which would not contribute to the global efforts to cut emissions. Lawsuits seeking to apportion responsibility for climatic events have generally failed in the past. Looking ahead, lawyers believe that a new branch of legal arguments called 'Attribution Science' should allow the courts to decide with reasonable confidence that these individual climate-induced extreme weather events have been exacerbated by planned human activities and industrialisation. Governments and firms could then risk being sued in future, if they fail to manage climate change and cut their carbon emissions.

Oceans climate change impacts and outlooks

Global warming and climate changes have led to serious impacts on oceans around the whole planet. These serious climate-induced ocean impacts included higher sea temperatures, rising sea levels, increased sea acidity and worsening marine life ecosystems. These climate-induced impacts are having severe implications for sea-life and aquatic ecosystems in various oceans globally.

Climate change and global warming have led to serious increases in sea levels around the world. These have resulted from the accelerated melting of the polar ice caps plus the thermal expansions of seawater due to the higher sea and atmospheric temperatures. The average sea levels around the world have risen by about 8 inches or 20 cm over the past 100 years. Satellite data have shown that there has been an average increase in global sea levels of some 3mm per year in recent decades. A large proportion of the change in sea level has been accounted for by the thermal expansion of seawater due to global warming. As seawater warms up due to global warming, the molecules become less densely packed, which then causes increases in the volume of the oceans around the earth (PNAS, 2018).

The accelerated melting of mountain glaciers and the retreat of polar ice sheets have also contributed to sea level rises around the world. Most glaciers in the temperate regions and along the Antarctic Peninsula have been melting and in retreat due to global warming. Since 1979, satellite records had shown dramatic declines in Arctic sea-ice sheet coverages.

Looking ahead to the next 100 years, climate scientists globally are fore-casting that the sea levels globally would rise more rapidly due to global warming and climate changes. Coastal cities, such as New York, Hong Kong, etc., have already seen increased flooding risks. Looking ahead to 2050, many coastal cities would likely require new seawall defences to help them to survive. Conservatively sea levels have been forecasted to rise by 1 to 4 feet, or 30 to 100 centimetres globally with global warming. These fore-cast sea level rises would be enough to flood many coastal regions, including small Pacific island states such as Vanuatu. Many famous beach resorts and coastal cities such as Bangkok, Boston and Hilton Head would suffer higher numbers of serious flooding incidents.

If global warming and climate change further worsen in future, then in the worst scenario the Greenland ice cap and the Antarctic ice shelf could suffer accelerated melting and eventually disintegrate. Then the sea levels globally could rise by as much as 20 feet or 6 metres, which is 5 times the forecasted normal sea level rises. These would then result in major flooding incidents which would inundate large parts of Florida, the Gulf Coast, New Orleans and Houston in the US plus many other coastal cities in Europe, Asia, Africa and Middle East regions.

The UN Intergovernmental Panel on Climate Change (UN IPCC) has also forecasted that sea levels might rise by 40 centimetres over the next 100 years conservatively. Many climate and ocean scientists have forecasted that future sea level rises could be much higher leading to flooding in many coastal areas. A serious climate-induced flooding example is Bangladesh, which has suffered many severe flooding incidents every year, affecting millions of its citizens. UK cities such as London, Bournemouth, Cardiff, Newcastle, Carlisle and Edinburgh could also face serious flooding risks (UN IPCC, 2017).

Climate change and global warming have also seriously affected the acid-ity of the oceans. Over the past millions of years, the earth's oceans have maintained a relatively stable acidity level. These steady ocean conditions have supported the development and growth of the rich marine life in the oceans around the world. Research by ocean scientists has shown that the delicate ocean balances have been endangered by the recent rapid drops in ocean surface pH caused by climate change. These would have devastating consequences and impacts on fish and marine life in various oceans globally.

Climate and ocean scientists have shown the higher CO_2 emissions, since the beginning of the Industrial Revolution, have led to serious acidity changes in the oceans around the world. Industrialisation with rising use of fossil fuel-powered machines and combustions has resulted in higher emis-sion of CO_2 and other GHGs into the earth's atmosphere. Scientists have shown that about a third to half of this anthropogenic, or man-made, CO_2 would be absorbed over time by the oceans around the world. The CO_2 ab-sorption by oceans has helped to slow down the climate change impacts that these high CO_2 emissions would have instigated if they had remained in the atmosphere. Recent ocean research results have shown that the absorption of these massive amounts of CO_2 by the seas around the world has been altering the ocean water chemistry and making the oceans more acidic. These pH changes have seriously affected the life cycles of many marine organisms, particularly those at the lower end of the food chain.

The oceans around the world have been playing very important roles in carbon absorption and storage. Researchers have shown that the oceans globally have been absorbing about 22 to 25 million tons of CO_2 per day. When CO_2 is absorbed and dissolves in the oceans then carbonic acid is formed. This then leads to higher seawater acidity, mainly near the surface. These acidity changes have been shown to inhibit sea shell growths in marine animals and to cause reproductive disorders in fish species in the oceans around the world.

Over the last 300 million years, the pH levels of the world's oceans have remained slightly basic with an averaged pH value of about 8.2. Recent ocean measurements have shown that the pH of oceans have been reduced to around 8.1, which is a drop of a 0.1 pH unit. This 25 per cent increase in acidity over the past two centuries in the oceans around the world has been caused by climate change. The higher GHG CO_2 emissions have led to higher atmospheric CO_2 concentrations, which have in turn resulted in higher CO_2 absorption by the oceans.

Looking ahead to 2100, scientific projections show that the continued rising CO_2 emissions could reduce ocean pH by another 0.5 units, which is 5 times the reduction to date. These higher acidity levels could seriously affect the life of many shell-forming sea animals including corals, oysters, shrimps, lobsters plus many planktonic organisms and some fish species. The more acidic seawater would seriously damage the ability of sea creatures to form shells which are essential for their wellbeing. The marine shell species have been the basis of the ocean food pyramid. Hence their loss would then seriously threaten the ocean food chains which would affect the wellbeing of many marine species.

It is important to recognise that whilst the oceans have been continuing to absorb more CO_2, their overall capacity for carbon capture and storage would likely diminish in the future. This would then result in more of the CO_2 emissions remaining in the earth's atmosphere, which would further aggravate global warming and climate change.

The scientific researches into ocean acidification and its implications have started relatively recently. There is lots more work required to study in depth its effects and impacts on various marine ecosystems. All the research has shown that urgent action would be required to reduce fossil fuel CO_2 emissions. Otherwise the rising ocean acidification would threaten many ocean organisms and marine ecosystems. This could lead to large numbers of marine species perishing or dying off unless they are able to evolve quickly to accommodate the serious ocean condition changes induced by climate changes.

Latest ocean research results have shown that the world's oceans have been heating up more than previously believed. Ocean scientists, using the latest ocean temperature measurement robotic technologies, have shown that there has not been any hiatus in ocean temperature rises globally. The latest ocean temperature measurements from 2014 have given much more precise estimates of ocean warming and temperature rises. The more accurate ocean temperature results have been generated by a new fleet of ocean monitoring robots called Argo. These robotic fleets comprise nearly 4,000 floating robots which have been launched to drift throughout the world's oceans globally. These Argo robots have dived, every few days, to a depth of 2,000 metres to measure the ocean's temperature, pH, salinity and collect other relevant ocean information. These Argo ocean robot measurements have helped to provide consistent and up-to-date data on ocean heat content since the mid-2000s. The latest robust ocean measurement evidence showed that oceans have been warming more rapidly than previously thought. The latest ocean results showed that about 93 per cent of the excess heat energies, trapped by GHGs globally, have been accumulating in the world's oceans, which have resulted in higher ocean temperature rises. These new ocean measurement results have raised serious concerns about the rising pace of climate change and their serious impacts on the different oceans around the world (Princeton University, 2018).

Coral reef climate change impacts and outlooks

Climate change and global warming are two serious threats to coral reef ecosystems globally. The higher ocean temperatures and ocean acidification caused by climate change and global warming have seriously altered ocean chemistry and have negative impacts on corals globally. These have caused serious damages to the coral reef communities globally with widespread coral bleaching, rising infectious diseases plus the death of many coral species globally. These negative climate-induced coral impacts have also seriously affected many sea organisms and fish that use coral reefs as their habitat.

Climate change impacts have led to oceans and seawater becoming warmer. The warmer seawater temperatures have put serious stresses on coral reefs because corals are very sensitive to changes in temperature. Warmer seawater temperatures have caused wide spread coral bleaching and rising infectious diseases globally. When the seawater is too warm, the corals will expel the algae or zooxanthellae that are living in their tissues. These have led to widespread coral bleaching with coral reefs turning completely white. Without zooxanthellae, the corals will turn white because zooxanthellae give corals their colour. Coral reefs which are bleached white are unhealthy corals which are weaker and less able to combat diseases. Bleaching events on coral reefs have been reported in various oceans globally. Sad examples of serious coral bleaching caused by climate change have been reported in the Pacific islands plus the National Park of American Samoa.

Looking ahead, as climate change worsens, oceans would become even warmer than today. Looking ahead to 2050, global mean surface temperatures have been forecasted to further increase by some 1.9 to 2.6 degrees Celsius. Looking ahead to 2100, the global mean surface temperatures have been forecasted to rise even further by a range of 3.1 to 5.2 degrees Celsius. These warmer sea temperatures would cause even more coral damages and bleaching incidents around the world (WWF, 2019).

In addition to ocean warming, climate change and rising CO_2 emissions have also caused serious ocean acidification which has serious negative impacts on coral and sea life. The higher CO_2 emissions have led to higher CO_2 atmospheric concentrations. These have then resulted in higher amounts of CO_2 being absorbed by the ocean which resulted in serious changes to the seawater pH and chemistry. Scientists have shown that the oceans have been absorbing about a third of the CO_2 produced since the Industrial Revolution and about half of all the CO_2 produced by burning fossil fuels on earth. The oceans do not have infinite capacities to absorb CO_2 emissions. Looking ahead, the oceans are likely to be less able to absorb more CO_2 as their capacities become limited.

The rising ocean acidification has led to coral reefs being less able to absorb the calcium carbonate they need to maintain their skeletons. This has led the stony skeletons supporting corals reefs to gradually dissolve and eventually disintegrate. Ocean scientists have shown that ocean acidification has lowered the pH of the ocean by about 0.11 units which have reduced the ocean's pH from 8.2 to a lower pH of 8.1 to date. This relatively small change has meant that the oceans globally have become about 30 per cent more acidic now than they were in 1750. These negative climate-induced seawater pH changes have seriously reduced the coral calcification rates. These in turn have negatively affected the coral reef building capacities, leading to extensive coral reef breakdowns and coral damage. Furthermore, many sea organisms and fishes associated with coral reefs have also been negatively affected. These negative impacts highlight the need for urgent actions to control climate change and ocean acidification so as to protect coral and marine ecosystems.

Coral scientist researches have shown that the rising ocean acidification rates could result in reductions of about a third of the coral calcification rates. This would mean that the amount of calcium carbonate absorbed by coral from the seawater would be reduced by 33 per cent, which is a big reduction. These reductions would not lead directly to coral death in the way that bleaching would, but it would impair the ability of coral to grow, to repair and to reproduce. In the wake of coral bleaching caused by higher seawater temperatures, corals will be less able to recover if the ocean's pH continues to drop. Looking ahead to 2060, ocean scientists have forecasted that the ocean pH could fall further to 8.0 from the current seawater pH of 8.1. Before the start of the Industrial Revolution, which led to accelerated GHG emissions, the oceans had a pH of 8.2. The 0.1 to 0.2 unit seawater pH drop might seem to be relatively minor changes but leading marine scientists globally have forecasted that there would be significant negative impacts on the coral ecosystem globally. Looking ahead to 2100, if CO_2 continues to be produced at the current rates, then experts have predicted that future atmospheric CO_2 increases would be high enough to lower ocean surface pH to 7.8 by 2100. Studies have indicated that these new lower seawater pH levels could dissolve many coral skeletons, leading to extensive disintegrations of coral reefs in oceans globally (Albright et al., 2016).

Climate changes and ocean acidification have also seriously affected the healthy growths of coral polyps globally. Healthy coral polyps need to absorb calcium carbonate from the ocean water to build their skeletons. The healthy growth of coral polyps has already been seriously impacted by the climate-induced ocean acidification to date.

So, looking ahead, the warmer seawater temperatures and rising ocean acidity levels will likely cause more coral bleaching and damages globally. These will seriously affect the health of many coral reefs globally as the bleached corals will become less able to combat diseases. If extensive coral reefs are lost in various oceans globally then this will cause vital habitat to be lost to many sea life and fish. Hence it is urgent to control climate change and reduce CO_2 emissions globally so as to ensure sustainable developments of coral reefs and the many associated marine organisms.

In addition, ocean acidification has affected more than just corals in the various oceans globally. Other valuable sea-life species, including snails, clams and urchins, which also make calcium carbonate shells, have also been negatively affected by ocean acidification. Just like corals, ocean acidification has made it much harder for these sea organisms to absorb the calcium carbonate which they need to build their shells from the seawater.

Climate changes, in addition to ocean warming and acidification, have also led to global sea level rises and altered ocean circulation patterns. Sea level rises have caused increases in ocean sedimentation which in turn have led to extensive smothering of coral reefs globally. In addition, changes in storm patterns by climate change have led to stronger and more frequent storms which have caused extensive damages to corals globally. In the worst cases, severe storms have caused widespread destruction of coral reefs in different oceans. Industrialisation and heavier storms have also led to increased land-based runoff of contaminated water, sediment and land pollutions from coastal regions into the surrounding seas. These have had major negative impacts on corals and sea life globally. These land-based pollution sources include coastal development, deforestation, agricultural runoff plus oil and chemical spills etc. These have all seriously impeded coral growths and reproduction plus disrupted overall ecological functions. In addition, these pollutions have also caused diseases and raised the mortality in sensitive coral and marine species. Scientists have shown that many serious coral reef ecosystem stressors have originated from land-based sources. These have included toxicants, sediments, and pollutants etc. The climate-induced, land-based pollutant runoffs from coastal cities into the surrounding seas have also contributed to extensive algal blooms. These have resulted in murky water conditions which reduced the available lights in the seas which would then negatively affect coral growths. Altered ocean currents have led to serious changes in connectivity and temperature regimes. These have resulted in serious reduction of food availabilities for corals and hampered the dispersal of coral larvae in the oceans.

Extensive coral damage examples could be found within the USA, where coral reef ecosystems at various locations have been highly impacted by watershed alteration, runoff, and coastal development. On the US islands in the Pacific and Caribbean, there have been serious coral damages resulting from climate change and pollution. Many of these pollution issues have been made worse by the specific geographic and climatic characteristics found in these tropical island areas. The combined effects of climate change and pollution have caused serious damage to coral reefs and their ecosystems globally. The negative impacts include coral bleaching, lower coral calcification rates, worsening coral integrity and death of coral reefs, and so on.

All these climate-induced coral damages have in turn negatively affected the marine food supply chains that coral reef ecosystems have been providing to coastal communities in different oceans. The coral reef ecosystems globally have been supporting important commercial, recreational and subsistence human fishery activities globally. Fishing has been playing a central social and cultural role in many island and coastal communities. Fishing has provided critical sources of food and also income for many coastal communities, especially in developing economies. The unsustainable high levels of fishing in coral reef areas have led to coral damage by fishing nets and the depletion of key coral reef species in many locations. These coral losses could often have wide ripple effects. These would not only just affect the coral reef ecosystems themselves, but also the local economies that have depended on coral. Additionally, many fishing nets and gears have inflicted serious physical damage to coral reefs, seagrass beds, and other important marine habitats. Rapid human population growth, increased use of more efficient fishery technologies, plus inadequate management and enforcement have led to the depletion of key reef species and habitat damages in many ocean locations.

So overall, climate change has caused serious damage to coral reefs and their ecosystems globally. Looking ahead, climate change would lead to further ocean temperature rises and ocean acidification plus rising sedimentation and pollution. These would result in further serious damage to coral reefs and their ecosystems around the world. Hence it is urgent to control climate change and reduce CO_2 emissions globally so as to ensure the future healthy developments of coral reefs and associated marine organisms.

Ice and glacier climate change impacts

Polar scientists have shown that climate change effects have incurred serious damages on the world's ice glaciers and polar ice sheets. Analysis of the global ttemperatures records going back to the late 19th Century have shown that the average temperature of the earth's surface has increased by about 0.8 degrees Celsius (1.4 degrees Fahrenheit) in the last 100 years. It is alarming that the bulk of this global earth surface temperature rise, about 0.6 degrees Celsius (1.0 degree Fahrenheit), has occurred in the last three decades. These serious global warmings have led to many glaciers in the temperate regions of the world and along the Antarctic Peninsula melting and suffering serious retreats.

Satellite records have shown, since 1979, a dramatic decline in the Arctic sea-ice coverage areas. There has been an alarming ice sheet reduction rate of some 4 per cent per decade. In 2012, the ice sheets coverage areas reached alarming low values which were 50 per cent lower than those in 1979–2000. The Greenland ice sheets have also experienced record melting in recent years. Polar scientists estimated that if the entire 2.8 million cu km of the Greenland ice sheet were to melt, due to global warming, then it could raise sea levels by six metres globally. These would then lead to serious flooding of many coastal cities and islands. Satellite data has also shown that the West Antarctic ice sheet has also suffered accelerated melting and has been losing mass. A recent polar study indicated that the East Antarctica ice sheet, which had so far displayed no clear warming or cooling trends, might also have started to melt and has been losing mass in the last few years.

Another good evidence of Arctic ice melting is that polar scientists have reported that their monitoring of the Arctic sea ice levels has shown that the annual low points in the summer seasons have now gone below what would have been a yearly low in the 1980s. This means that current Arctic sea ice areas, when compared to the average areas from the 1980s, 1990s and 2000s, have fallen below the annual low levels recorded in the 1980s. This dramatic ice sheet melting and reduction has raised serious concerns globally. It has demonstrated clearly that there has been long-term melting of the Arctic and Antarctic polar ice sheets. This has been caused by global warming plus serious GHG emissions globally (NSIDC, 2018).

The serious melting of the polar ice sheets caused by global warming has also generated other serious secondary climate change impacts globally. One of the most serious climate-induced impacts has been the continued rise in sea levels globally. Satellite data has shown that there has been an average annual increase in the global sea levels of some 3mm per year in the recent decades. Global warming has also made the changes in sea levels worse due to the thermal expansion of seawater as a result of higher sea temperatures. As the seawater warms up due to global warming, the seawater molecules will become less densely packed which then will cause an increase in the volume of the ocean. So global warming has caused the melting of mountain glaciers and the retreat of polar ice sheets plus sea level rises globally.

Furthermore the Arctic ice melting has generated major impacts on various weather patterns around the world. The Arctic sea ice sheets have helped to maintain a fine balance on global warming in the past by reflecting the incoming solar sun rays back to space thus helping to balance and regulate the earth's temperatures. Industrialisation and rising human activities have released more GHGs into the atmosphere. The subsequent global warming and temperature rises have accelerated the melting of the Polar ice sheets. This melting has resulted in more of the Arctic ocean areas becoming exposed after the loss of ice coverage. The more exposed sea surfaces have absorbed more solar energy, which has then raised the sea temperatures further. This has resulted in a vicious cycle of heat absorption and sea warming which could then lead to further accelerated melting of the polar ice sheets.

In addition, more and more of the Arctic and Antarctic polar areas have been opened up recently for commercial activities, including oil and gas exploration, commercial shipping etc., by different countries globally. The higher level of industrial and commercial activities could generate additional serious environmental impacts, leading to more accelerated ice melting plus more damage to the fragile polar ecosystems.

Recent Arctic temperature research has shown that temperatures in the Arctic have been rising at twice the rates of global temperature rises when compared to the rest of the earth as a whole. This has demonstrated that global warming has been having more pronounced serious impacts on the Arctic polar regions than the rest of the world. The resultant Arctic ice loss has also caused serious negative impacts to the fragile local polar ecosystems. These have resulted in important Arctic animals, such as walruses, penguins and polar bears, losing their critical habitats. The reduced polar food chain supply has also threatened the continued survival of important polar species.

A good example of the serious climate impacts on the Arctic ecosystem is on the king penguins. Polar scientists have shown that king penguins will be under serious survival threat if nothing is done to constrain the negative climate change impacts on the polar region. Scientists have measured the penguins' fragmented population in the Southern Arctic ocean regions. They have found that some key polar island penguin nesting sites would soon become unsustainable in future due to climate change. Scientists have reported that almost 70 per cent of the king penguins, which is about 1.1 million breeding pairs, would have to be relocated by 2100. If they could not be relocated successfully then they would risk dying before the end of the century because of global warming and GHG emissions (CNRS, 2017).

Climate change and global warming have caused the available food supplies for the penguins to become more and more distant for them to reach. They have to travel longer distances to fetch sufficient food for their young chicks on their traditional polar nesting sites. Scientists have found that the penguins would need a good supply of fish and squid which have normally been found in the Southern Ocean Antarctic Polar Front (APF). This is a nutrient-rich region which is currently providing abundant food for the hungry penguin and their young chicks. Global warming has caused the APF to move further south poleward. If the global temperatures continue to rise then this valuable APF region would likely move out of the range of many foraging king penguins in future. Scientists have found that seven hundred kilometres is about the maximum limit at which the penguins could travel and not risk exposing their chicks to starvation back at their nests. In particular the Marion and Prince Edward Islands, and Crozet Island, which have been major penguin nesting sites, would be experiencing the biggest difficulty in the next 50 years. If climate change worsens even further, then many penguin sites including those nesting sites on Kerguelen, Falkland and Tierra del Fuego Islands would run into major difficulties. Many king penguins are likely to face death and many of these sites could become extinct, if climate change and global warming continue.

Looking ahead to 2100, continued global warming would likely cause more accelerated melting of the polar ice caps and glacier around the world. In the worst climate change scenario, scientists have forecasted that the world's glaciers could disappear gradually over the next 100 years. In addition, there would also be accelerated melting of the huge Greenland ice sheet, the Arctic polar ice caps and the Antarctic ice shelf. Continued rising global warming would result in them melting quickly and reducing in size significantly. These would then lead to other serious climate-induced impacts globally, including rising sea levels plus more flooding of coastal cities and islands globally.

Global warming has already caused snowfalls to decline at some of the world's most popular ski resorts, etc. A good example is that many skiing operators at top ski resorts around the world have already been require to apply more artificial snow generation to ensure that there would be good snow coverage on their ski runs. The natural snow falls have diminished to unacceptable low levels in recent years due to global warming and climate change.

So overall, climate change and global warming have caused serious damage to ice and glaciers globally. In addition, more countries opening up more of the Arctic and Antarctic polar regions for resource exploration plus commercial shipping would likely cause more serious damage. This would likely lead to serious environmental damage to the fragile polar ecosystems plus accelerated melting of the polar ice sheets and dying of some valuable polar species. Hence there are urgent needs to jointly reduce climate change and minimise global warming so as to arrest the melting of the polar ice caps and glaciers globally so as to protect the fragile polar ecosystems.

Ecosystems and agriculture climate change impacts

Global warming and climate change have resulted in serious impacts on many ecosystems globally plus negatively affected agricultural crops and plants globally. The rising global temperatures have also forced many insect, bird, fish and animal species to have to migrate to cooler regions which have more suitable habitats for them. These climate- induced migrations have generated serious changes in the earth's fragile ecosystems plus affected the growth of agricultural crops globally. In addition, these have also serious additional implications on local economic development, employments and businesses. Details of these will be discussed below with international examples.

A good example of climate-induced ecosystem impact is that global warming and rising sea temperatures have pushed many fish species to have to migrate over longer sea distances to go to cooler oceans where the seawater will be at the right lower temperatures. These climate-enforced fish migrations have also led to serious social and economic challenges to many coastal fishing communities, especially in the developing countries. The fish migrations have led to poorer catches for the fishermen in the warmer regions. In some worst cases, they have eliminated the fishing livelihood for some coastal fishing villages when the catches became too low. In many fishing ports, such as those on the US east coast, the fishing fleets have to sail further out to sea to reach their new fishing grounds.

Climate change has also resulted in serious impacts on plants and wildlife globally. Many species of plants and animals have been negatively affected by climate change. A good example is that as the polar ice sheets have continued to melt due to global warming, these have led to serious sufferings for the polar seals and polar bears. Many polar animals have died due to reduced food availability and suitable polar habitats disappearing. The fine balances in the polar ecosystems have been seriously upset by climate change and global warming. A good example is that if the polar seals were to die out then it would mean less food for the polar bears and other polar animals who would then have to struggle hard to survive. So climate change could also lead to widespread starvation and death of many important polar animals.

Climate change and global warming have also seriously impacted the microscopic plankton and the tiny krill in the polar seawater. These plankton and krill are key parts of the polar sea food supply chain and have been providing vital food supplies for a huge number of polar marine species. These have included barnacles, fish, sharks and whales. Plankton and krill are very sensitive to sea temperatures changes. They have been dying and reducing in numbers as the sea temperatures have risen as a result of global warming. This has led to significant reductions in the amount of food supplies available for polar fish and marine species, forcing them to struggle for survival.

Climate change and global warming have also affected many insects, such as mosquitoes and egrets. Global warming has led to rising temperatures in many countries which have forced these insects to spread into parts of the world that were previously too cold. A good example of this is that the egret used to be quite rare in the UK but now it can be seen regularly in large numbers in various river estuaries in the south of England. Many of the faster moving animals, such as birds, have been migrating to other areas as conditions in their habitat have changed for the worse due to climate change. Sadly, the slower moving animals, like snails and frogs, have not fared so well because they could not move away as easily. Many plants have also not coped well with climate change, as they could not move and evolve fast enough. Looking ahead, as global warming and climate change continues, weather conditions could be changing faster than many plants, insects and animals are able to adapt. So climate-induced changes could eventually cause some plants, insects and animal species to die and become extinct.

Climate change and global warming have also led to changes in the weather systems in many regions globally. These would include changes in temperature, rainfall and wind patterns. These weather changes have resulted in drastic ecological changes in many regions globally. A significant example is that in California, climate-induced changes have led to more serious forest fires, which have caused serious damage to local communities, especially the fires around Los Angeles recently.

Climate change and global warming have also had serious impacts on agriculture, farming and food production globally. In fact, some of the most striking implications of climate change have been felt in the agricultural sectors globally. These climate-induced impacts have been felt differently in various regions around the world. The largely temperate developed countries have been experiencing different climate-induced changes versus the more tropical developing countries. Agricultural crops normally grow best at quite specific temperatures and climatic conditions. Rising temperatures and climate- induced changes have had serious negative impacts on crop productivity globally. A good climate-induced crop problem example is that the productivity of rice, which has been the staple food of more than one third of the world's population, is very sensitive to temperature rises. Rice productivity will generally decline by 10 per cent for every 1 degree Celsius increase in temperature. Global warming has seriously affected rice productivity in the hotter traditional rice growing regions around the world. The rising temperatures have pushed staple crops such as rice away from their traditional growing fields around the equator further north to the cooler temperate regions. These climate-induced rice migrations have significant negative impacts on the economy and livelihood of rural farming communities in the various traditional rice-growing countries.

Another good climate-induced farming example is that in North America, the rising temperatures have led to reduced corn and wheat productivity in the US mid-west. These regions have traditionally been the main producer for corn and wheat in the USA. On the other hand, the rising temperatures in the cooler northern parts of the America continent have led to rising crop productions in the northern USA and in Canada.

Climate changes have also been changing the weather systems and rainfall in different regions. These changes have been having serious implications on agriculture and farming in different regions. A good agricultural example is that farmers in the temperate zones have found that the drier hotter conditions have made it more difficult to grow crops such as corn and wheat. This has led to the once prime agricultural growing regions facing serious threats to their continued economic developments in future.

Farmers have been able to handle some of these climate-induced problems by applying new crop technology and higher applications of fertilizer. Continued innovations in crop technology including GM and chemical fertilizer developments would be required to tackle the growing climate change impacts on agriculture. However these could also led to other pollution and sustainability issues.

Rising global temperatures have also promoted the growth of agricultural pests and the wider spread of crop diseases. The agricultural pest populations have been on the rise globally due to global warming and climate change. For example, locusts have been spreading into areas where they have never been found before. Aphids or greenfly have also been hatching earlier in the year due to global warming. As a result they have been eating away at the young, delicate seedlings of valuable food crops. Crop diseases that were once found only in tropical areas have now become more endemic in much wider areas globally (Carbon Brief, 2018).

Climate experts have predicted that there will be major changes in the weather of different countries due to global warming in the future. For example, climate scientists have predicted that the British climate would become much warmer, like that of the Loire valley of France by 2060. As a result, farming experts have predicted that this could meant that the crops of sunflowers and oil seed rape, which could be used for cooking oil and cattle feed, would become more popular in future. In addition, vineyards for growing grapes for eating and for making wine would become more popular in the southern and central regions of the UK in future. Global warming and climate change could lead these regions to enjoy a warmer and dryer climate in future, like the climate in southern France now.

Looking ahead to 2050, agricultural scientists have forecasted that global warming and future temperature rises could seriously reduce rice production by up to 25 per cent by mid-20th century globally. At the same time, global population growth models have suggested that the population in the developing economies globally would rise quicker than the population of developed economies. By 2050, the global population would likely have increased by another 3 billion people. As a result, the developing world food producers would need to double their staple food crop production so as to maintain adequate levels of food supplies for their growing population. The reduced crop productivities caused by climate change would then put serious strains on the various food supply chains globally. These would lead to future food consumption pattern shifts globally and in the worst cases, rising starvation in some poor regions.

So overall climate change and global warming have caused major changes in many crop ecosystems and affected crop productivities. These climate-induced changes have serious impacts on the livelihoods of farmers and sustainable economic developments of the traditional crop growing countries. There is urgent need for joint actions to control climate change and global warming so as to minimise these negative agricultural impacts (EEA, 2018).

Human health climate impacts and outlooks

Leading scientists and doctors have reported that there have been serious negative climate-induced impacts on human health and major health risks from the combined effects of climate change, global warming and pollution. These serious negative human implications include pollution, disease, shortening of life, etc. Climate change and industrialisation have led to serious air pollution in many cities globally. Air pollution has become a major public health crisis globally due to the rising pollutant emissions and high particulates emissions. Much of the air pollution has been generated by the energy and power generation sectors with fossil fuel combustion. Scientists have shown that some 3 million deaths annually could be attributed to poor air quality and pollutions globally. These have made climate-induced pollutions to become the world's fourth-largest threat to human health. The other three top causes include high blood pressure, dietary risks and smoking. Looking ahead, experts have predicted that climate-induced pollution would take further toll on human life and health globally. The number of global deaths due to climate-induced pollution would rise even further globally (IEA, World Energy Outlook, 2017).

Scientists have investigated and demonstrated the close linkages between climate change, fossil energy, air pollution and human health. Fossil fuels used for energy production, especially those from poorly regulated or inefficient fossil fuel combustion, have been shown to be the single most important man-made sources of air pollutant emissions globally. It has been estimated that some 85 per cent of particulate matter and almost all of the sulphur oxides and nitrogen oxides have been generated from these fossil fuel sources. These three pollutants have been responsible for the most widespread impacts of air pollution globally. Scientists have estimated that this climate-induced pollution has caused more than 3 million annual deaths globally (IEA, World Energy Outlook, 2017).

Scientists have forecasted that the number of climate- and pollution-induced deaths would rise further in the future. The rising regional demographic trends plus increased energy usages and urbanisation, especially in developing countries in Asia, would mean that the number of premature deaths attributable to outdoor air pollution would continue to grow. Experts have forecasted that climate-induced air pollution deaths would likely rise by 150 per cent from 3 million today to 4.5 million by 2040. It is particularly worrying that emerging economies in Asia will most likely be accounting for almost 90 per cent of the rise in premature deaths. The key reason is that the rising air pollution in many of the emerging economy cities in Asia will create serious public health hazards to their populations and will be shortening their citizens' expected life spans significantly.

Climate change-induced air pollution has particularly adverse impacts on the aging population and elders globally. This is particularly worrying for many countries with aging populations, such as China, Japan, USA and

Europe. For example, the aging population in China have been seriously affected by air pollution, with increased occurrence of lung cancer and respiratory diseases. These have been particularly serious in Northern China where the air pollutions have been particularly serious. With new government policies to reduce pollution and for cleaner energy applications, the air qualities have been slowly improving in China. Despite the decline in aggregate pollutant emissions in some regions, the lasting health damage generated by past serious pollution incidents will be hard to reverse. These have led to serious illnesses and poor health, especially for the weaker and more elderly population. These have resulted in high hospitalisation and medical costs. A particularly serious example is the rapid rise in lung cancer cases for non-smokers in many of the heavily polluted cities in China and Asia.

Household air pollutions have also been shown to be major causes of death globally. Scientists estimated that currently about 3.5 million deaths per year have been caused by household air pollutions. These have been caused by uncontrolled fossil fuel combustions in some households in developing countries. These have included the uncontrolled use of dirty old coal burners or oil burners in rural areas. In some extreme cases, the practice of burning waste plastic as fuel in some poor coastal fishing villages has also resulted in higher incidents of cancer and the early death of villagers. Burning plastic releases carbon monoxide, dioxins and furans which are some of the most toxic chemicals known to science and human (WECF, 2005).

The new stricter climate change and cleaner energy policies being introduced by various Governments globally should help to control and improve air pollutions globally. These are much needed improvements but the air qualities in many cities will take time to improve and the health damages will be hard to reverse quickly. A good example of household pollution improvements is the provision of improved cooking stoves and alternatives to solid biomass for rural communities. These have been part of the new policies of improving access to clean cooking facilities in rural areas in different emerging economies. Looking ahead to 2040, the number of people without access to clean cooking facilities globally is projected to fall by almost 1 to 1.8 billion from the 2.8 billion today. As a result, the number of premature deaths attributable each year due to household air pollution has been forecasted to fall from around 3.5 million today to under 3 million by 2040.

In addition, climate change and global warming have led to warmer winters, which have been encouraging more disease growth and spread. The key reason is that warmer winters are now not cold enough to kill off many nasty germs and bacteria. This has then resulted in the growth and multiplication of germs and nasty diseases globally, such as flu and measles, etc. Increased global temperature rises have also increased the reproduction rates of microbes and insects. These have also contributed to the accelerated growth and spread of diseases. In addition, various disease microbes have been evolving faster and speeding up the rate at which they can develop resistance to control measures and drugs. These further add to the health problems.

A serious climate-induced disease example is that mosquitoes carrying the malaria disease used to be found only in hot tropical countries. Global warming has resulted in the malaria carrying mosquitoes spreading further north to the warmer temperate regions. There are serious fears that malaria could soon reach Britain and Europe, where it could become a major health risk, as it has been in the tropically regions. Another serious disease is dengue fever, which was previously largely confined to tropical areas but have recently become endemic in wider regions due to global warming.

Climate change and global warming has also led to hotter summers globally especially in Europe and Americas. These have generated adverse impacts on human health and have caused some deaths. A serious climate-induced health example is that there have been many more heat-related deaths plus rising cases of heatstroke and dehydration in Europe in recent years. These have occurred as summer temperatures have risen abnormally high levels, above 35 to 40 degrees Celsius, due to global warming.

Medical research have also shown that climate change and environmental pollution have caused serious impacts on human health, especially on lung and heart. Medical researchers in leading UK universities have demonstrated the serious impacts by pollutants on heart and lung health. Their researches have demonstrated the negative impact of urban air pollution on cardiovascular and respiratory health. These have shown that even short-term exposure to serious air pollution in built-up areas in inner cities will generated very negative damaging impacts on the health of hearts and lungs (Imperial College London, Climate Change and Environmental Pollution Affects Heart and Lung Health, 2018).

Latest scientific reports have shown that air pollution could have killed even more people than previously estimated. Scientists estimated that nearly 8 to 9 million people across the world each year might have been killed or affected by air pollution. The World Health Organization (WHO) has previously estimated air pollution was to blame for up to 4.5 million deaths across the world. WHO has also previously estimated that tobacco smoking was responsible for 7.2 million deaths globally in 2015. However medical researchers have recently reviewed the health data to show that the true pollution toll could be over 8 million deaths. Most of the additional deaths have been caused by pollution-induced heart diseases. These show that air pollution has led to more deaths than smoking globally (Bell, 2019).

To overcome these adverse health impacts caused by climate change and air pollution, urgent joint actions by governments and energy companies will be required. The International Energy Agency (IEA) has proposed three key areas for urgent joint actions globally (IEA, WEO Special Air Pollution Report, 2017).

Firstly, there are urgent needs to establish joint ambitious long-term air quality goals globally. Different key stakeholders should subscribe to these new goals. In addition, the efficacy of various pollution mitigation options should be assessed against these new air quality targets globally.

Secondly, there are urgent needs to put in place a package of clean air policies for the energy sector to achieve with long-term goals. These new policies should include a cost-effective mix of direct emissions controls and new clean energy policy objectives.

Thirdly, there should be effective international monitoring, enforcement and evaluation so as to keep the pollution improvement strategies and measures on course. These would require reliable air quality data and measurements. There should also be strong and continuous focus on compliance by various governments and companies to these new targets. In addition, there should be continuous efforts on policy improvements plus timely and transparent public information.

IEA has forecasted that if these new clean air strategies were to be implemented well in different countries, then these should help to cut pollutant emissions by more than half in future. To achieve these improvements, it would require various governments and energy companies to take joint actions together. In addition, they have to co-ordinate effectively with each other, so as to deliver comprehensive overall improvements to air pollution and climate change globally (IEA, WEO Special Air Pollution Report, 2017).

Migration and refugee climate impacts

Climate change and global warming have created serious migration and refugee problems globally. The recent World Bank refugee report have highlighted the various negative impacts of climate changes which have resulted in major migration and refugee issues in Africa, Asia and Latin America. Looking ahead, they forecasted that if climate change worsens further then this could lead to 140 million climate refugees by 2050 globally (World Bank, Refugee Population by Country Data, 2018).

Historically people have migrated between different regions globally for various reasons, including social, economic or political drivers. Climate change has now emerged to be one of most important major drivers of migrations globally. The various negative impacts of climate change, including droughts, flooding, failing crops and severe storms, etc., have contributed to higher migration and refugees, especially in the Africa and Asia regions. These negative climate change incidents have propelled increasing numbers of people to move from the affected regions which have become inhospitable, to the more viable and hospitable areas of their countries or other continents so as to rebuild their lives. The World Bank have analysed migration and refugee trends on three regions including Sub-Saharan Africa, South Asia, and Latin America. They have estimated that these three regions together represented 55 per cent of the developing world's population. By 2050, experts forecasted that international climate migrations could rise further unless there are significant cuts in GHG emissions plus robust development actions. Experts forecasted that the worsening impacts of climate change on these three densely populated regions of the world, i.e. Sub-Saharan Africa,

South Asia, and Latin America, could lead to more than 140 million people, or more than 2.8 per cent of the population, being forced to move within their own countries so as to escape the negative impacts of climate change by 2050. These climate-induced migrants and refugees would be moving away from the less viable inhospitable areas with lower water availability and poor crop productivity. Some migrants would have to move from coastal areas affected by rising sea level and storm surges due to climate change. These major new climate-induced migrations and refugee movements would have major impacts on the various countries involved. The refugees could impose huge strains on the local infrastructures, hospitals and social support systems of the various countries (World Bank, Groundswell: Preparing for Internal Climate Migration, 2018).

There are urgent needs for concerted joint actions globally to control climate change so as to lower refugee migrations. These would include joint global efforts to cut GHG emissions together with robust development planning at the various country levels. Experts forecasted that effective joint global efforts to cut GHG emissions together with robust development planning at the country levels could dramatically reduce the worst-case refugee migration scenarios by as much as 80 per cent, or 100 million fewer migrants. Experts have warned that there is only a small window now for joint action, before the effects of climate change worsen to a dangerous tipping point.

Looking ahead, developed countries and cities globally would also need to make preparations to cope with the expected upward trends of migrants and refugees arriving from the less developed countries and rural areas. There should be preparations for improving infrastructures, schools, hospitals and housing etc. In addition, there should be preparations to create suitable opportunities for education, training and jobs for the refugees. It would also be important to help refugees to make better decisions about whether to stay where they are or take the dangerous journey to move to new locations. There should also be sufficient help for the refugees on their long journeys, often with young children and the elderly.

Climate and migration experts have looked closely at migration in three specific countries, Ethiopia, Bangladesh and Mexico. These are three countries with very different climatic, livelihood, demographic, migration and development patterns. Their specific climate refugee problems are discussed in more detail below.

In Bangladesh, the villages have been suffering serious floods every year with climate change. The village houses have often been badly affected and damaged by the floods. In addition, many rice paddies have been washed away, resulting in many village farmers losing their livelihoods. So many villagers have been leaving their villages to move to the capital city, Dhaka. Some refugees have been able to be connected to the NARI project. This is a World Bank initiative designed to provide training, transitional housing, counselling and job placement services for poor and vulnerable women. These programs have helped them to support their family back in the villages, which highlights the importance of good development planning and refugee support programs. They are essential to help countries to be better prepared for their rising future climate refugee migrations.

Looking ahead, experts have forecasted that there might be up to 40 million internal climate migrants in South Asia by 2050. Bangladesh has been forecasted to be contributing a third of these climate refugees, with some 13 million climate refugees. Currently close to half of Bangladesh's population have been dependent on farming and agriculture. Bangladesh has been undertaking various initiatives in the water, health, forestry, agriculture and infrastructure sectors so as to incorporate climate adaptation into its national development plans. Several adaptation programs have been undertaken in Bangladesh, including a program to enhance food security in the north-west of the country plus another program which encouraged labour migration from the north-west during the dry season.

In Africa, experts forecasted that Sub-Saharan Africa could have more than 85 million internal climate migrants by 2050. Ethiopia would be one of the most vulnerable countries to climate change in Africa, due to its reliance on rain-fed agriculture. Ethiopia's population is also likely to grow by 60 to 85 per cent by 2050. This would then place severe additional pressures on Ethiopia's natural resources and infrastructures. Ethiopia has been taking steps to diversify its economy and to prepare for its increasing internal migrations, with inputs from international agencies including the UN.

In Latin America, experts have forecasted that there might be over 15 million internal climate-induced migrants by 2050. Mexico is a large and diverse country. Its rain-fed farming and crop growing areas would likely experience the greatest 'out-migration'. This would be caused by declining crop productivity, which would lead to increased hardship and poverty. Global warming would also increase in the average and extreme temperatures, especially in the low-lying and hotter regions, such as coastal Mexico and especially the Yucatan. The poorest populations in Mexico would be worst affected. These would include the climate-sensitive smallholders, self-employed farmers and independent farmers. All of these social groups would also have higher than average poverty rates and be very susceptible to negative climate impacts.

For the future, it is very important that leading governments and countries globally should all recognise that internal climate migration would be a growing reality and burden for many countries. To prevent these migrations escalating to crisis levels, the governments of the affected countries should start to introduce relevant new improvement policies so as to reduce and control the serious rising migration trends. Experts have forecasted that these countries could potentially reduce the number of people forced to move due to climate change by as much as 80 per cent by 2050 by taking actions in three main areas.

Firstly, it is vitally important for the countries to cut GHG emissions now with new improved climate change and clean energy policies. Strong global climate actions would be needed to meet the Paris Agreement's goal of limiting future temperature increases to less than 2 degrees Celsius by 2100. However, even at the reduced levels of global warming, some countries would still face high levels of internal climate migration by 2050.

Secondly, it would be important for governments to embed climate migrations in their development planning. Countries should integrate climate migration planning into their national development plans. Most regions currently are poorly equipped to deal with the rising migrations from rural areas facing increasing climate risks, into urban areas which might already be heavily populated. To secure good resilience and development prospects, suitable preparations and planning would be required.

Thirdly, countries should start investing now to improve their refugee data collection and analysis, especially on the scale and scope of local climate migrations. More investments would be needed to better understand the likely scale and magnitude of climate-induced migrations in future. Good country-level migration and refugee modelling would be important for good policy formulation and development planning.

Researchers have also found that that over 80 per cent of the refugees displaced by climate change have been women. The 2015 Paris Agreement has made specific provisions for the empowerment of women as it was recognized that they have been disproportionately impacted by climate change. There has been serious displacement of women by climate-induced incidents in Nigeria, Cameroon, Chad, Niger and the Central African Republic.

For example, in Central Africa up to 90 per cent of Lake Chad has disappeared due to climate change. As the lake's shorelines have receded, women have to walk much further to collect water. The women's roles as the primary caregivers and providers of food and fuel in these countries have made them more vulnerable to serious floods and droughts resulting from global warming. These situations have been caused by the fact that in many of these African countries, most of the men have to go into the various townships for work during the dry months. As a result, they have to leave the women behind to look after their children and the community. With climate change causing the dry seasons to become much longer, the women have to work harder to feed and care for their families without support. In recognition of these problems, governments and organisations working on climate change have started to include specific provisions for women refugees in their policy formulation and development planning (BBC, Climate Change 'Impacts Women More than Men', 2018).

Climate change damages and cost impacts

Climate change and global warming have led to various serious extreme weather events around the world. These serious climate-induced events have negatively affected thousands of lives in various countries and caused huge damage throughout the world. Scientists have identified that there have been

10 extreme weather events in 2018 around the world which have resulted in damages costing more than $1bn each. In addition, there have been 4 major extreme weather events which have caused damages costing more than $7 billion each (Christian Aid, 2018).

Climate scientists have also analysed the causes of each of these extreme weather events. They have found that the increased occurrence of extreme heat waves in Europe was influenced directly by human-related global warming activities. They have also found that the other extreme weather events have resulted from shifts in global weather patterns which were consequences of climate change and global warming.

The analysis showed that the most financially costly weather disasters in 2018 linked to global warming were Hurricanes Florence and Michael. It was estimated that the costs of damages from Hurricane Florence was around $17 billion. The costs of damage for Hurricane Michael were estimated to be $15 billion. Climate researchers have shown that the heavy rainfalls accompanying Hurricane Florence were made 50 per cent worse due to human influenced global warming effects. With Hurricane Michael, experts said global warming had caused the warming of the seawater which added fuel to the storm, making the wind speeds stronger. All the negative climate effects have made these extreme weather events very damaging.

In 2018, Japan also suffered an extreme summer, with flooding and heatwaves which caused huge damages. The flooding in Japan persisted for weeks and killed more than 230 people. It was estimated that the flooding caused over $7 billion worth of damage. After the flooding, Japan was hit by Typhoon Jebi, which was one of the most powerful storms to hit Japan for 25 years.

In 2018, Europe experienced record heatwaves in various European countries. These have led to negative impacts on many communities with big cost impacts and deaths. Climate researchers showed that climate change and global warming effects have most likely doubled the chances of these extreme heatwave events happening in Europe in recent years.

In the UK, an independent study from the UK Met Office suggested that the extreme heatwave events that UK experienced in 2018 were made 30 times more likely because of the rising global warming resulting from climate change. They have also predicted that the UK summer temperatures could rise by more than 5 degrees Celsius by 2070, whilst winter temperatures could also rise by up to 4 degrees Celsius. Their climate model projections have shown that the UK's average yearly temperature could be more than 2 degrees Celsius higher by 2100. In addition, the average summer rainfall could also decrease by up to 47 per cent by 2070, whilst there could be up to a 35 per cent chance of having more rain and precipitation in winter. Experts have also predicted an increase in both the frequency and magnitude of extreme water levels around the UK coastline. In particular, the sea levels in London could rise by more than 1 metre by 2100 due to global warming and melting of the Arctic ice sheets. These would increase flooding risks for major areas of London, including the financial centres in the City of London plus the Houses of Parliament in Westminster (UK Met Office, 2018).

Global temperature monitoring has shown that 2018 was the fourth warmest on record. The earth's average temperature was close to 1 degree Celsius above the levels recorded in 1850–1900. In addition, twenty of the warmest years on record have also taken place in the past 22 years due to climate change and global warming (WMO, 2018).

Looking ahead, experts have forecasted that there would be high likelihoods of further global temperature rises in 2019 which might lead to a new El Niño globally. Experts have predicted that the global average temperature for 2019 could be 1.1 degree Celsius, above the pre-industrial average period from 1850–1900 (UK Met Office, 2018).

In addition, climate scientists also believe that global warming has been driving major shifts in weather patterns on the earth that have made heatwave, hurricanes, droughts and wildfires much more likely than before. Climate scientists have undertaken more attribution studies to link the extreme weather events to climate change and global warming. They have predicted that the combined effects of climate change and global warming would likely cause more extreme weather events globally, including record temperature rises, heatwaves, droughts, hurricane etc. Climate researches have shown that for many people, climate change and global warming are having devastating impacts on their lives and livelihoods right now. The unprecedented floods, droughts, heatwaves, wildfires and super-storms that have been experienced in various countries globally have been shown to be linked to climate change and global warming. Hence it is important for different countries across the world to work together to manage climate change and to minimise global warming. Key joint global climate change mitigation actions would include reducing carbon emissions, increased usage of clean energy and renewables, improving energy efficiencies, promoting clean transport and electric vehicles etc. These will be discussed in more detail in the following chapters.

3 Climate change global policies management

求人不如求己
Qiú rén bù rú qiú jǐ
It is better to rely on yourself rather than depend on the help from others.
If you want a thing done well, do it yourself.

Executive overview

Many governments and countries around the world have realised the importance of working together to better manage climate change and to reduce global warming. The Paris Agreement was a good milestone. The COP24 meeting in 2018 successfully developed the Paris Rulebook. There are still many difficulties in developing suitable new climate and energy policies plus achieving effective joint climate change management globally. Key hurdles include some governments and companies considering that climate actions could incur extra costs which would affect their short-term economic growths and business performances. Details of different new climate and energy policies developments by different key countries and regions across the world will be discussed further in this chapter.

Climate change government policy overviews and outlooks

Many governments and countries around the world have recognised the importance of working together on controlling climate change and reducing global warming. In practice, there have been many difficulties in achieving effective joint climate change management globally. Key hurdles include that some policy makers and companies have viewed climate management to be expensive plus that new climate change measures would require major changes in policies and life styles in different countries globally. Some executives in the traditional business sectors, especially those in the utilities and fossil fuel sectors, have viewed climate change measures as incurring additional costs for their businesses which would reduce their short-term profitability.

A serious climate business example would be for utilities and power generation companies to retrofit their fossil fuel power station with new flue gas and waste treatment facilities so as to meet the new stringent emission requirements introduced by various governments. Many have been reluctant to make these extra investments as these retrofit costs could reduce their short-term profits. However they should realise that these new climate improvements will be part of their license to operate sustainably in future.

Globally many governments have been developing and enacting new clean energy policies plus carbon management systems including carbon emission trading or carbon tax systems. These should reduce GHG emissions and promote businesses to actively incorporate climate change as part of their business planning. A good carbon tax example is the new Carbon Tax system that is being introduced by the Federal Government of Canada recently. They have introduced a new carbon tax system which is in line with their new climate change policies and their Paris Agreement commitments. The Federal and Provincial governments of Canada have used the revenues generate by their new carbon tax to fund various climate change improvements across Canada such as funding for their new green public transport systems in Toronto (Alini, 2019).

The United Nations Framework Convention on Climate Change (UNFCCC) has been organising the global effort to manage climate change by the different countries. The UNFCCC was launched at the 1992 Rio Earth Summit. It has aspired to reduce the global greenhouse gas GHG emissions to a lower level which would prevent dangerous anthropogenic interference with the climate system globally. It has initially set some voluntary GHG emissions reduction targets which the various countries globally have not been able to meet. After the failure of the Rio initiatives, the 191 signatories to the UNFCCC then agreed to meet in Kyoto in 1997 to establish a more stringent climate change regime. The resulting Kyoto Protocol created a global carbon credits trading system with binding GHG reductions for the various ratifying countries. Some key countries were not part of this Kyoto Protocol. Key country examples include the USA, which did not sign the Kyoto Protocol, plus China and India, which were both exempt as developing countries (UNFCCC, 2012).

In follow-up to the Kyoto Protocol, the annual Conferences of the Parties (COPs) to the Kyoto Protocol have been held almost every year. The COP meetings have been held in different locations such as The Hague, Cancun and Doha. However little progress has been made, mainly due to the serious conflicting interests between the various countries. In 2012 the Kyoto carbon trading system was not renewed and it collapsed after the failure of the 2012 Doha COP meetings.

In 2016, the various countries met in the COP21 meetings in Paris. After long, intense and difficult negotiations, the various countries managed to made major advances in their negotiations. Finally many countries were able to jointly negotiate and agree the Paris Agreement. At the end of the UN climate conference in Paris in December 2015, 195 nations unanimously agreed to restrict the global temperature rise to less than 2 degrees Celsius, or preferably 1.5 degrees Celsius, which is above the pre-industrial 'baseline'. This was an important political achievement and agreement by the 195 countries. However, to achieve the very stretched and ambitious targets set, it would require almost a complete cessation of global CO_2 emissions by the second half of this century. This will be very challenging to achieve and will involve very difficult discussions by the various countries to agree on how to achieved these challenging targets. In the meantime, the world has crossed the important 400ppm mark in CO_2 concentrations globally, which has led to a global temperature rise of 1 degree Celsius (UNFCCC, Paris Agreement, 2016).

The Paris Agreement also reaffirmed a commitment to reduce emissions by the 195 different countries to achieve a 2 degrees Celsius global warming target. These would include mandatory reductions by developed countries plus calls upon developing countries to contribute. They also created an international climate change fund which would be used to compensate climate change losers. They have also re-established a new Kyoto style Clean Development Mechanism (CDM) and carbon trading system. It is important to note that the Paris Agreement has made no provisions for agricultural emission reductions and also largely ignored the developing economies. These are important areas that would also have big implications for climate change and global warming. More difficult negotiations would need to be undertaken by the various countries so as to reach agreements in these two important areas.

The Paris Agreement has anticipated that future revisions and refinements would be required every five years. These would provide the much-needed scope for incremental policy reviews and developments at the national country levels. A practical approach to climate change policymaking should complement the Paris framework. This would provide flexibility for individual governments to facilitate the use of more incremental and adaptable policy responses. These should better reflect the local resource endowments and socio-economic circumstances.

The 195 countries involved in the Paris Agreement have been working on developing their detailed individual country plans to meet their climate commitments. Looking ahead, there will be many challenges and uncertainties to manage. These would include serious issues such as the USA not ratifying the Paris Agreement, plus the various uncertainties about how different countries would be implementing their agreed commitments.

A good climate example is that the International Energy Authority IEA has shown that the energy-related CO_2 emissions have seriously contributed to the large majority of global greenhouse gas (GHG) emissions to date. As a result, many countries have been introducing new clean energy policies to transform their energy mixes so as to reduce their greenhouse gas emissions as part of their fight against climate change. However, the implications have been challenging for most countries. Even if all the countries were able to meet their emission goals as pledged under the UNFCCC, it would still leave the world with 13.7 billion tonnes of CO_2 emissions. This would be equivalent to 60 per cent above the levels that would be required to remain on track to meet just 2°C global warming by 2035. Details of these various serious climate change challenges and various potential new opportunities will be discussed in more detail in the sections below plus various other chapters of this book.

COP24 2018 meeting highlights and implications

The COP24 meeting held in Poland in December 2018 had over 23,000 attendees from different countries globally. There was important sharing of information between the participants as well as tough negotiations. There were some concerns amongst many representatives that some leading countries have not been placing sufficient priority on climate change. A good example is that 200 parliamentarians from different countries had an important meeting together during a weekend in the middle of the COP24. They discussed their common concerns that, following the IPCC 1.5 degree Celsius report, many governments globally have still not done enough to reduce their greenhouse gas emissions. They discussed important climate change topics including how the parliaments of various countries can adopt appropriate legal frameworks for their countries to meet their Paris Agreement commitments. In addition, they discussed how different parliaments globally could hold their governments and politicians accountable for implementing the climate mitigation actions agreed in the Paris Agreement. Recent research has shown that less than 10 per cent of the countries globally have implemented domestic legal frameworks which would be compatible with their Nationally Determined Contributions (NDCs) (LSE Grantham Institute, 2018).

After tough negotiations during COP24, the Paris Rulebook was finally agreed and delivered. At COP21 constructive ambiguity was used to allow countries with different views to sign up to the Paris Agreement. To reach agreement on the new Rulebook required elimination of these ambiguities between all the countries. There were challenging negotiations taking into account the different technicalities and political positions globally. The agreed Rulebook that was finally agreed and adopted by all the countries comprised 133 pages plus additions and appendixes. The agreed Paris Rulebook should help to provide a common reporting structure for all countries on global climate adaptation and mitigation. This should help to build trust between countries and promote climate actions globally. It should also help various countries to plan, implement and review their climate adaptation and mitigation actions. It might encourage some countries to ratchet up their climate change ambitions and actions.

The agreed Rulebook contained detailed guidelines on the planning for climate adaptation and actions. It also provided better improved ex-ante and ex-post information on the climate finances that have been pledged. It covered transparency on future climate finances. However it was difficult for developed countries to commit to finance beyond their budget cycles. These have often been from one year to a maximum of four years. The Rulebook also helped to strengthen environmental integrity and emission reductions reporting. The Rulebook included specific climate and emission reporting guidelines which have been based on strict scientific basis. These should help to reduce the use of various accounting tricks that have occurred in the past.

There was one significant area which was not agreed at COP24. This was on the important Article 6 on global carbon markets. The key reason was that some countries wanted a more permissive carbon market regime than other countries. This could potentially lead to problems on double carbon accounting. In addition, there was also some serious technical work outstanding. Hence it was agreed by the delegates at COP24 to postpone the carbon markets discussions by another year. The delay was preferred by all delegates rather than forcing agreements on some weak carbon market rules.

In addition, the climate finance ministerial dialogue was good. Many of the Less Developed Countries (LDC) will need capacity support to improve their climate finance reporting. There has also not been enough sharing and capacity building to the local levels and on adaptation. The climate finance package could have been stronger and it was agreed to postpone setting up a new goal for climate finance to 2020.

One encouraging development was that most of the countries, apart from a few oil producing countries, have all agreed that it would be important to focus on meeting the new IPCC target of 1.5 degree Celsius target rather than the current 2.0 to 3.5 degree Centigrade envelopes. This was a major global megatrend shift following the publication of the IPCC 1.5 degree Celsius report. The IPCC report highlighted the negative impacts of climate change globally and stressed the importance of urgent joint climate change actions internationally. It was disappointing that some countries, especially the oil and gas producing countries, have chosen not to welcome and adopt the IPCC 1.5C report due to their own specific national interests. These hurdles will be major challenges for future COP meetings to deal with and to resolve.

Following the agreement of the Rulebook, countries globally will now need to focus on improving their NDCs to get to the 1.5 degree Celsius envelope rather than the previous 2.0 to 3.5 degree Celsius envelope. The common playing field established by the Rulebook should help to make countries more willing to consider these improvements. In addition, most countries recognised that their BAU would not be sufficient to deliver the required climate actions. Looking ahead, the countries would need to discuss their improved NDC in the 2019 climate summit. Then they have to meet the key requirement to deliver their new NDC updates at the planned COP meeting in 2020.

It was recognised that, as countries introduce new policies to reduce their emissions, these could potentially raise social license and public acceptance issues in the future. These problems have led to serious public protests and demonstrations in some countries. One recent example was the Yellow Vests protests in France and demonstrations in Paris about the fuel price increases proposed by the French government. It would be important for governments globally to realise that their new climate and clean energy policies could disproportionately affect some social segments in their country, especially the lower income families which might suffer some social hardships. Hence the appropriate Climate Change Just Transition process planning and actions should be included as an integral part of various countries' overall climate action plans.

The agreed Rulebook would be common for all countries globally. It also provides enough flexibility for developing countries taking into account their special national conditions and development status. During COP24, the LDC pushed for a strong rulebook taking into account their special situation and requirements. The LDC was comprised of 47 of the poorest countries in the world. Their total population amounted to 1 billion people globally but they accounted for less than 1 per cent of the GHG emissions globally. These countries have already been badly impacted by various climate change impacts, including flooding, drought and plastic wastes. The LDC countries have been developing new long-term initiatives for effective climate adaptation and resilience. They are planning to finalise these initiatives by 2020 so that they can present these at the COP meeting in 2020.

In summary, COP24 was a partial technical success but there would be more work required to build momentum around raising national climate ambitions and mitigation actions. Whilst governments have managed to agree a lengthy Paris Rulebook, they could not yet finalise future rules for carbon markets and climate finance which they have agreed to postpone for decision by one year. More importantly, most governments have to step up their national climate targets and mitigation actions. The findings of the IPCC's 1.5C report have shown that greenhouse gas emissions have actually risen recently after being stable for 3 years. The rules for reviewing progress against the Paris Agreement and raising ambition in the 2020s have to be further developed including serious national reviews with clear and transparent guidelines globally.

Post-COP24 climate actions and implications

COP24 was crucial in providing the Paris Rulebook which should help to provide a more level playing field on climate reporting globally. It also increased the understanding of the need to raise climate targets and ambitions for different countries. Looking ahead, COP25 and COP26 will be critical meetings when countries should lead on ramping up their climate targets for 2030. These would be not easy tasks as there would be big hurdles to overcome by different countries. In the next 20 years, the economies of the world globally are expected to double with the planned economic growths in both developed and emerging economies. At the same time, the world would need to reduce carbon emissions by over 30 per cent so as to achieve the 2 degree Celsius target. To achieve the 1.5 degree Centigrade target then the world would need to reduce carbon emissions by over 50%. To put these into perspective, the annual economic growth rates are expected to be around 3.5% but the decarbonisation rate for 2C will need to be 5.1%. For the IPCCC 1.5C target, the required decarbonisation rate will be a massive 6.7% (Stern, 2018).

Following the publication of the IPCC 1.5C report in 2018 and tough negotiations in COP24, general agreements were reached at COP24 that different countries around the world would need to take more actions to try to achieve the 1.5C targets. Many stakeholders have not clearly understood the possible existential risks plus the immense tasks which would face all countries when they start to roll out their tough new low carbon policies. In addition, the benefits of the low-carbon economy transitions, which could generate significant clean energy growths, have also not been clear to many people around the world.

On Climate Finance, the US$100 billion that was proposed in Copenhagen as being all public money, has been changed to both public and private in the latest negotiations. The public element in the US$100 billion fund would be key. Concessionary finance would be needed as well as more work to leverage contributions from the private sector. There would be much more work to do but the process is slowly moving ahead.

On the new carbon markets mechanisms, it was agreed in COP24 that it would be best to postpone decision on these for another year. The richer developed countries would like to claim more credit for providing overseas aids rather than gaining carbon offsets and reducing their own ambitions. The poorer less developed countries have been looking at developing new national carbon initiatives and schemes. It is important to recognise that carbon trading would only reduce the cost of initial emissions cuts. All countries would need to strive to achieve carbon neutrality by 2050 if the world would aspire to achieve the 1.5 degrees Celsius targets, as proposed in the IPCC 1.5 degrees Celsius report.

The New Climate Economy Report (NCE) 2018 has shown how countries at all levels of economic developments and income could embrace the shift to low carbon economy. This should help them to unlock inclusive growths in their countries in the 21st century. The new investments in sustainable infrastructure, innovations and new discoveries should help to drive new sustainable inclusive economic growths across different countries. At the same time, it is important to realise that there would be no new high carbon growth stories globally in future. These would be unsustainable and would eventually self-destruct due to the hostile environmental impacts created by these high carbon projects (Global Commission on the Economy and Climate, 2018).

Climate Change Just Transition will become more important in different countries globally. It is about social justice plus how people and communities globally would be affected by major changes in the global economy and climate changes. It is important to recognise that the shifts in the global economy would not just be about economic transition to a low carbon economy. It should also cover difficult transitions involving the social, industrial and services sectors. In addition, new developments in Artificial Intelligence (AI) and robotics could lead to major changes in employment and labour saving developments. A lot of these changes would be generated by the digital transformations and digital economic transitions which have been happening globally. We would need to become better at managing these difficult transitions globally under the Just Transition process. These would require life-long learning, bringing in new skills and universities plus improving finance and training for SMEs and businesses together. Countries will need to develop suitable plans to handle dislocations, which could include unemployment and workforce retraining plus demands for labour with advance digital skills. As part of the Just Transition processes, suitable safety nets will need to be developed but hopefully the number of people who will need these could be minimised by good planning.

The new climate economy and just transition developments should help different countries across the world to get more understanding of the requirements and difficulties in the climate transition. There is a long way to go in many areas, plus many stakeholders haven't taken on board the magnitude of the tasks and the required ramp-ups. More work will need to be done to realise the desired inclusive growths for clean energy and low carbon economy developments.

Global climate policy developments and challenges

The historic global climate treaty, the Paris Agreement, was signed by many countries globally in late 2015 after difficult negotiations at the COP21 meeting in Paris. Three years after the signing of the Paris Agreement, many countries have been developing their new climate policies in line with their Paris Agreement commitments. Some encouraging post–Paris Agreement developments included new renewables targets being adopted in different countries globally. A good example is that at COP22, the leaders of 48 developing nations committed to work towards achieving 100 per cent renewable energy in their respective nations. Another good example is that following the signing of the Paris Agreement, 117 countries have submitted their first NDCs as required under the Paris Agreement, and 55 of these countries have featured new renewable energy targets.

However some countries have also been reluctant to ratify the Paris Agreement and to meet their commitments. In some countries climate developments have been stalling or regressing in some cases. In some developed countries, where the use of fossil fuels has created comfortable lifestyles, their commitments on the fight against climate change and global warming have faltered recently with the election of new governments, such as the USA. On the other hand, some leading developing countries, including India and China, have emerged as new global leaders in tackling climate change and global warming. During the same periods, many countries have also experienced climate-induced extreme weather events which have led to heavy damages and costs. These included record heatwaves in Europe and America, plus hurricanes in the Americas, droughts in Africa plus flooding in Asia etc. (Climate Central, 2017).

The development of climate policies in various key countries and regions globally will be discussed in more detail below, with international examples.

American region climate policy overviews

In the United States, there have been significant reversals and backtracking on their climate policies. Since President Trump's inauguration, the USA has gone from being a champion of global climate actions to seriously considering not ratifying the Paris Agreement. The new Administration has also been pushing to end or relax various environmental and climate change regulations introduced by previous US presidents. A good example is that US President Trump has been trying to dismantle various environmental regulations as well as reversing President Obama's push for domestic and global solutions to global warming. The new US Administration has also been moving to stop funding for various climate change work and research. The US Government has been reducing spending on global climate programs in addition to rolling back regulations that might threaten the fossil fuel sector. There have been uncertainties over future US climate positions and policy directions. However, there is considerable opposition in the USA to these new directions from climate change supporters. In some Democratic-run states in the USA, environmental groups have been challenging President Trump's de-regulation drive with public campaigns and demonstrations. There have also been lawsuits launched in some US law courts. The Republican Party has lost its slim majorities in Congress in 2018 and this will introduce more future uncertainties. Some Republican lawmakers have also been voicing their support for climate actions and adopting different positions against their president. There are considerable uncertainties on how the US Clean Air Act would be amended or how the US Federal Government would reduce greenhouse gases emissions in line with their Paris Agreement commitments. At the recent COP24 climate meeting, the US State Department negotiators, despite President Trump's expressed views on the Paris Agreement, have been actively participating in the international climate change negotiations. It seemed that their negotiation strategy was to push for a new global deal on the new Paris Rulebook accord which this US administration and future US administrations, could support in the possible scenario that the US does not exit the Paris Agreement or rejoin in future.

In Canada there has been a significant shift in the Federal Government's climate positions, starting in late 2015 with the election of the Liberal government. The Canadian Government has changed from refusing to act meaningfully on climate change warming to becoming a strong advocate for climate change, after the government changed from the conservative to the liberal party. Their former conservative Prime Minister Stephen Harper had been reluctant to act meaningfully on climate change or develop appropriate policies. The new liberal Prime Minister Justin Trudeau has become a strong advocate for climate actions after winning the national election. Trudeau has moved to expand the climate programs ran by provinces that charge fees on climate pollution into a new nationwide carbon tax system. A new federal budget was introduced that included billions of dollars in planned spending on clean energy and climate programs. Canada has also been moving to new nationwide carbon pricings so as to reduce its greenhouse gas emissions. They are aspiring to become economically competitive in a low carbon global economy. They have been developing a new national carbon pricing policy plus developing targeted investments and policy directed at stimulating clean innovation. However there have also been deeply mixed messages about the future of Canada's tar sands oil industry in view of the potential social and unemployment problems. There have been past statements about the phasing out of Canada's tar sands oil mining industry. However Prime Minister Trudeau has recently said during a speech at an energy industry event that the Canada tar sands resources should be developed responsibly, safely and sustainably.

In South America there have been varying degrees of commitments to climate change by the different countries. Some countries have promoted renewable clean energy, such as Costa Rica. In other countries, there have been significant concerns about their commitment to climate change and global warming, especially deforestation. There are particularly serious concerns that deforestation has accelerated in both 2015 and 2016 following a decade of gains. Brazil had been initially successful in slowing down deforestation in their Amazon rainforest for the past decade by preventing the conversion of forest into agricultural land for beef and soy. Brazil has now reported that there have been spikes in forest loss in both 2015 and 2016. In 2015, Amazon deforestation increased by about a quarter compared with 2014. Deforestation further spiked by more than a quarter in 2016. The World Resources Institute had said that the recent rises in Amazonian deforestation have been caused by lax law enforcement of illegal wood logging. In nearby Paraguay, there has also been serious deforestation of a drier type of tropical forest that was not considered a part of the Amazon forest.

Europe and Russia climate policy overviews

The European Union (EU) countries have been some of the first developed economies in the world to take global warming and climate change seriously. They have announced their commitment and support for climate change actions. In recent years, the EU has been floundering in some of their commitment to further climate actions. Many EU countries have been facing pressures from economic slowdowns and refugee problems. Some EU countries have faced internal challenges, with a surge in opposition and far-right parties. There have also been rising public opposition and antipathy toward climate actions. A serious example is the yellow vest protests in Paris on fossil fuel price rises. The EU has already committed to reducing their greenhouse gas pollution by 40 percent by 2030, compared with their 1990 levels. These reductions are ambitious but are similar to a commitment made by the US state of California. Looking ahead, climate experts are expecting that the EU would be sticking to its objective to reduce emissions in 2030 by 40 per cent. However, the EU would be unlikely to go beyond these reductions as there has been rising opposition from Poland, Hungary and other countries.

In Russia there are strong concerns about their commitments to climate change and global warming. After their declaration that climate change was a global crisis, there have been uncertainties about the continued commitment of the Russian administration to climate change. Russia has been one of the world's biggest greenhouse gas emitters with its huge fossil oil and gas industries. The Russian economy continues to be heavily dependent on oil and gas fossil fuels sales to Europe, China and other countries. Russia appeared to be prepared to ratify the Paris Agreement. However Russia's pledge under the agreement would not require it to take any meaningful steps to slow down global warming. In 2015, Putin said that climate change has become one of the gravest challenges that humanity is facing. After Trump won power in the USA and expressed his doubts on climate change, the Russia government appeared to have changed their position again. There are serious global concerns if both Russia and the USA retreat from their climate change commitments.

Asia climate policy overviews

There have been encouraging developments in climate change policies and actions in many countries in Asia, especially in China and India. However some other Asian countries, including Australia and Indonesia, have been retreating from their climate change commitments with their new administrations.

In China there have been positive developments in climate change policies and actions. China has been embracing climate change and the low carbon economy transition as part of their new national policy on Ecological

Civilisation. China's leaders have changed their positions on climate change in recent years. They have put in place new sustainable development policies which would help to reduce environmental pollution and climate change, including greenhouse gas emission reductions. These major policy shifts have been further amplified recently as China has come to view the new low carbon economy and clean renewable technology as major drivers for sustainable economic growths. A good example is that China plans to create 13 million clean renewable energy jobs by 2030 with their planned new renewables investments. In 2015 it overtook the U.S. as the largest market for electric vehicles. China has also been delivering on its Paris climate goals far quickly than planned. Looking ahead, China could be in a good position to boost its climate pledges, if its economic transition goes as planned.

In India there have been encouraging developments in climate change. The national and local governments have both been boosting efforts to deploy clean renewable energy applications across India. The national governments as well as several local authorities have been actively implementing their new climate and clean energy policies. A large part of the funding for their climate actions would come from their national and sub-national governments. In addition, international finance would also be used to boost these efforts. India has also been developing one of the world's most aggressive plans for installing solar panels. This has been part of their national effort to provide electricity to the millions of residents who currently do not have regular access to electricity across India. A potential concern is that India's ambitious clean power plans have been relying heavily on financing and aids from the developed countries. Some experts are concerned that India could be negatively jeopardized by the climate policy shifts in the USA and other developed economies.

In Australia, there have been serious concerns about the government's climate change commitments changing in recent years. In 2013, one of the first major actions by the Australian conservative party after it won power was to dump the carbon tax which had been helping to slow climate change and global warming in Australia. Since then, the conservative party has replaced their Prime Minister Tony Abbott with the more moderate Malcolm Turnbull. The change in leadership has done little to bolster climate policy. Turnbull has been pushing for federal subsidies for a coal mine near the Great Barrier Reef. Climate experts, including the Australian National University Centre for Climate Economics and Policy, have expressed concerns about the uncertainties on future Australia climate change policy directions.

In some Southeast Asia countries, especially Indonesia, there are serious concerns about their climate change commitments. The global demands for palm oil have led to rampant deforestation in Indonesia. Efforts by international companies, such as Unilever, Nestle, Mars, to remove palm oil produced through deforestation from their supply chains have not yet arrested the deforestation. Natural forests continued to be burned and cleared at astonishing rates to grow palms in Indonesia and other Southeast Asian countries. Land has also been cleared for timber logging throughout the region and to produce rubber in countries that included Cambodia. These deforestation activities have been taking place despite a pledge by the Indonesia government to reduce its greenhouse gas emissions by nearly a third by 2030. Indonesia is one of the world's biggest greenhouse gas polluters, largely because of deforestation and fossil fuel industries.

Africa climate change policy overviews

Africa is one of the regions globally that has been most vulnerable to the impacts of climate change. Most of Africa has already experienced temperature increases of some 0.7 degrees Celsius and with further temperature rises likely. Africa has been facing a wide range of climate impacts and extreme weather incidents, including drought and floods, decreased food production, spread of waterborne diseases and risk of malaria, and changes in natural ecosystems and loss of biodiversity. There have been serious concerns that deforestation has been accelerating in Africa's biggest tropical rainforest. Deforestation has long been a major problem in the swampy Congo Basin in Africa. It traverses a number of African countries and is home to one of the world's greatest expanses of carbon-storing tropical forest. Timber is being harvested and trees are being cleared for mines, plantations and grazing. These problems have recently been getting much worse, with vast new logging hotspots identified by satellite images analysis. Researchers have found that the rate of deforestation have more than doubled in the Democratic Republic of Congo recently (UN, 2018).

Governments in Africa have been working through a number of regional and global institutions to strengthen their responses to climate change. They have been coordinating their regional positions and national policies on climate change through the African Ministerial Conference on the Environment (AMCEN). The secretariat has been provided by the UNEP based in Nairobi. The New Partnership for Africa's Development (NEPAD) has helped to promote climate projects and actions. African countries have also been tapping support from a range of funds and institutions. These included the Special Climate Change Fund and the Least Developed Country Fund created under the United Nations Framework Convention on Climate Change (UNFCCC), the Adaptation Fund under the Kyoto Protocol, the Global Environment Facility, the World Bank, and other UN and intergovernmental organisations and programmes. African countries have also participated in the Clean Development Mechanism (CDM).

4 Climate change sustainable transport management

英雄所见略同

Yīng xióng suǒ jiàn lüè tóng

The views of different heroes are generally similar.

Great minds think alike.

Executive overview

The transport sector has become one of the largest consumers of fossil energy globally. It has been consuming over a quarter of the overall energy consumption globally but has also generated over one fifth of the energy-related GHG emissions internationally. Fossil oil fuels have been accounting for over 90 per cent of the final energy consumption in the transport sector globally. In follow-up to the historic climate agreement in Paris in December 2015, the international community have been focussing increased attention on decarbonisation of their transport sectors. The potential options include electric vehicles, gas vehicles, hydrogen vehicles, biofuel, public transport, bike sharing etc. We shall discuss these various transport options plus strategies to decarbonise the transport sector in this chapter, with international examples.

Global green transport clean energy transformation

Global energy demands in the transport sector have been rising continuously by some 2 per cent annually on average since 2005. The transportation sector has been accounting for over a quarter, some 28 per cent, of the overall energy consumption globally. More seriously, the transportation sector globally has been generating over one fifth, about 23 per cent, of the energy-related GHG emissions of the world. This has been mainly due to the fact that oil fuels, especially gasoline and diesel, have been the dominant transport fuels being used globally. Fossil oil fuels have been accounting for over 90 per cent of the final energy consumption in the transport sector globally to date (REN21, 2018).

In follow-up to the historic climate agreement in Paris in December 2015, the international community have been focussing increased attention on the decarbonisation of the transport sector. About 22 countries have so far submitted NDCs that referred specifically to renewable energy applications in the transport sector. Two countries, including the island of Niue and New Zealand, have linked Electric Vehicle (EV) growth to renewable energy developments in their NDCs.

During 2016, some governments, mostly in Europe, had begun looking at new medium- to long-term strategies to decarbonise their transport sector. These new country climate policy and strategies would often involve long-term structural changes which closely link the transport sector with the energy and electricity sectors. A good example is Germany's climate action plan, developed in 2016, which aimed to reduce emissions in the transport sector by 40–42 per cent by 2030, with a longer-term objective to fully decarbonise the transport sector in Germany.

There are various potential options to decarbonise the transport sector globally. These include gas vehicles, electric vehicles, hydrogen vehicles, biofuel, public transport, bike sharing, etc. The main focus of recent international climate and decarbonisation discussions has been on the electrification of road transport and EV developments. There is also now increasing attention being focused on developing good reliable clean electricity charging systems linked to clean renewable electricity supplies so as to further support EV developments in different countries.

Analysing the transport and charging sector in detail showed that there are some potential quick options for clean renewable energy growths in the current transport supply chain. These growth options could include the use of liquid biofuels, biofuels blended with conventional fuels, natural gas vehicles, etc. In addition, there should be major investment in new infrastructures and retail station networks that would support liquid and gaseous biofuels supplies plus the charging of EVs, which can use batteries or hydrogen produced by renewable electricity, etc.

Biofuels, which can be ethanol and biodiesel, have represented the vast majority of the renewable share of global energy demands for transport. These biofuels have been supplying around 4 per cent of the road transport fuel in selected countries globally. The global ethanol productions have remained stable in recent years. There have been some biofuel declines across Europe and in Brazil which have been offset by increases in the United States, China and India. The global biodiesel productions have been growing faster, with substantial increases in the United States and Indonesia. These biodiesel growths have been fuelled by supporting policies from their governments. There have also been strong public concerns and policy drives against diesel fossil fuels in many countries, especially on their GHG and particulate emissions. These could particularly affect the future growth of biodiesel globally.

The various technologies for producing, purifying and upgrading natural gas and biogas for use in transport have become relatively matured. The gas infrastructures and supply chain for clean gas vehicles based on natural gas fuels have been growing slowly but steadily internationally. A good example is the growth of liquefied petroleum gas (LPG) and biogas vehicles in China especially for taxi fleets in Beijing and other leading cities in China. There are several barriers to broader gas and biogas penetrations in the transport sector. These include the lack of regulations regarding access to natural gas grids plus the lack of national natural gas and biogas transport supply and retail infrastructures. The decentralised nature of biogas feedstocks and their comparatively high economic costs have also been affecting growth. Most of the biogas production for transport purposes has been concentrated in Europe and the United States.

The electrification of the transport sector has been rising in recent years with the EV growth globally. These growths have helped to expand the potential for greater integration of clean renewable energy in the form of electricity supply to trains, light rail, trams, and two and four-wheeled EVs. Looking ahead, further electrification of the transport sector globally should create more new markets for clean renewable energy applications. The growth of clean renewable energy in various national grids has also helped to promote growths in their share of the electrified transport charging systems. Some EV service providers and utilities companies have started to offer new provisions for charging electric vehicles with clean renewable electricity. A good example is the new customer service offerings by new electric car sharing companies in the United Kingdom and the Netherlands.

On a very limited scale, some car companies in several countries have been developing new EV prototypes that could use solar PV directly. Interesting examples include new passenger car prototypes in China and Japan plus solar-powered buses in Uganda. These are still in the early stages of development, with relatively limited applications globally.

The key barriers to electrification in the road transport sector have continued to include relatively high EV purchase costs plus limited driving ranges and short battery life of EVs. The lack of national charging infrastructures in various countries has also limited EV growth. In most developing countries, there are additional barriers to EV growth, with the lack of robust local electricity supply infrastructures and availability of electric charging retail stations. These have seriously reduced the attractiveness for consumers to buy EV and use electricity charging.

Looking ahead, many countries, including Germany, India, the Netherlands, UK and Norway, have begun seriously discussing the gradual phase-out of the internal combustion engines. Some countries have been discussing possible phasing out of diesel engines by 2020–2030 and petrol engines by 2030–2040. These plans will need extensive consultation and careful development to ensure public acceptance.

The aviation transport sector has accounted for just over one tenth, around 11 per cent, of the total energy used in transport globally. The International Civil Aviation Organization announced in 2017 a landmark agreement by 66 nations, which accounted for 86 per cent of the global aviation activity, that they have agreed to jointly mitigate GHG emissions in the aviation sector. They are planning to begin the first phase of the agreement in 2021. The new agreement will support the production and use of sustainable aviation fuels. These new fuels could include clean aviation fuels produced from biomass and different types of waste. The use of biofuel in aviation had moved from a theoretical concept to a business reality for a few airlines already in 2016. A number of significant agreements for the provision of aviation biofuels were signed in 2016. Some of these aviation fuel agreements are likely to be worth over US$ 1 billion. In addition, there has also been ongoing development work on new electric aeroplane prototypes for short-range electric flights plus new solar powered planes (IRENA, 2017).

The shipping transport sector globally has been consuming less than one tenth, around 7 per cent, of the total energy used in transport globally. The International Maritime Organization have also agreed to a 0.5 per cent sulphur cap by 2020. This new cap will have serious implications for the burning of heavy fuel oil in ships and big impacts on oil refineries globally. The new cap will contribute to GHG emission reductions in line with global climate change concerns. These caps should increase interest in clean shell fuel energy applications, such as liquefied natural gas (LNG) and renewable fuel applications, in ships. The integration of various renewable energy into shipping activities have not gone very fast in recent years. Looking ahead, hopefully there may now be more developments with the planned fuel oil sulphur cap in 2020. In theory clean renewable energy including wind and solar energy could be incorporated directly into shipping activities. For ship propulsion, the use of biofuels or other renewable based fuels including hydrogen, can also be considered. There are also some new development activities associated with the use of gaseous fuels for ships. A good example is the deployment of LNG-fuelled ships in Australia, which may also offer opportunities for biogas incorporation. There has also been some research into developing new wind energy-assist technologies for ships (*Seatrade Maritime News*, 2018).

The railway transport sector has accounted for around 2 per cent of the total energy used in the transport sector globally. The majority of railway trains globally, around 50–60 per cent, have been fuelled by oil products. Over one third, about 36 per cent, of railway trains have been powered by electricity. The share of clean renewable electricity in the total energy mix of the world's railways has increased 3 times, from 3.4 per cent in 1990 to nearly one tenth, around 9 per cent globally. Some countries have reached much higher penetrations on clean renewable electricity usage in their railways. A good example is that all the electric trains in the Netherlands have been converted to clean wind energy generated electricity. A few other railways have implemented new projects to generate their own renewable electricity. Good examples of these renewable power railway applications can be found in India and Morocco. They have installed new wind turbines on railway land and solar panels on railway stations. Chile has also announced the construction of new solar PV and wind farms to help to power their Santiago subway. The testing of new smart on-board dynamic energy management system in both intercity and urban trains has been taking place. These should help to improve railway energy efficiency plus help to manage and store variable renewable energy for use on trains.

Regional green transport clean energy developments

Road transport has accounted for 75 per cent of transport energy uses globally. Each region has their unique mix of renewable fuels, vehicle types and fuelling infrastructures. The various key trends in road transport fuel consumption for different regions across the world will be discussed in more detail below.

In Asia, there has been continuous growth in ethanol and biofuel production for transport fuel applications. China, India and Thailand have been leading the Asia region in biofuel productions. The production of biodiesel has also continued to rise, especially in Indonesia. On biogas, both China and India have established gas infrastructure into which biogas could be fed for transport uses. A good example is India's first biomethane-fuelled bus, which has started operating recently. India is also planning more biogas stations, biogas buses and routes across their country.

In China, EV sales have risen significantly with the government's new energy car program. China has become the largest EV global market for passenger EVs and market leader for electric two-wheel motorbike sales globally. China has also overtaken both the USA and Europe in EV cumulative sales and has become the largest global electric bus market, with over 173,000 plug-in electric buses operating on its roads. Experts have forecast China will likely account for more than 50 per cent of the global electric bus market by 2025. These rising EV numbers in China should help to reduce GHG emissions and achieve China's Paris Agreement commitments.

In Japan, EV sales have been declining recently but still accounted for about 8 per cent of the global market for passenger EVs. Japan has also published a blueprint for future hydrogen transportation developments. They are also planning to deploy new hydrogen buses for the Japan Olympics.

In Europe, the policy and public support for first-generation biofuels have continued to wane due in part to sustainability concerns. The rising interest in electric mobility and EV has also contributed to the recent declines in biofuel interests and investments. The total EU regional production of both ethanol and biodiesel has been declining across the EU region. There were some increases in ethanol production in some countries, including Hungary, Poland, the UK and Sweden. Biogas and biomethane productions have grown in the EU to gain more share of the transport fuels market. A good example is Sweden, which provided record shares of biomethane, over 70 per cent, in its supply of compressed natural gas (CNG) for its transport sector. Four of the world's five largest producers of biogas for vehicle fuel are now based in 4 EU countries including Germany, Sweden, Switzerland and the UK. The EU regional sales of EVs have also been increasing in recent years and have accounted for nearly one third, about 29 per cent, of global sales of passenger EVs. Norway has been leading the EU region in total EV sales. The Netherlands is second, with the UK third and France in fourth position. The Netherlands has also built, completed and operated the first solar controlled, bi-directional charging station for EVs in the world.

In Africa, the production of biofuel and ethanol has risen in recent years. There have been some early EV sales in South Africa and Morocco. Some biogas and biomethane road transport pilot projects have been launched in South Africa in recent years.

In North America, the USA has continued to be the largest producer of biofuels. The US government has enacted supportive agricultural policy for biofuels together with the US federal renewable fuel standards. The production of both ethanol and biodiesel has increased recently, which reversed recent declines. The USA is also one of the five largest producers of biogas for vehicle fuel worldwide. Renewable biogas has accounted for about 20–35 per cent of the gas fuel used in the transport sector. EV sales have also increased in the USA with rising consumer demands. Globally, the US EV sales have accounted for not quite one third, about 28 per cent, of the passenger EV sales in the global market. In Canada, the ethanol biofuel production decreased whilst biodiesel production increased. EV sales have also increased significantly with rising consumer demand.

In Latin America, Brazil has become the second largest producer of biofuels globally after the United States. However there have been some recent declines in both ethanol and biodiesel consumptions in Brazil. There have also been decreases in both ethanol and biodiesel productions in Colombia and Peru. On the other hand, the production of biofuels has increased in Argentina and in Mexico. The EV market in Latin America is still in its early development stages. Argentina, Brazil and Colombia have also developed natural gas infrastructures into which biogas could be incorporated.

Global green transport policy developments

Government policy supports to improve the sustainability of the transport sector have traditionally occurred in two key areas, which include improving energy efficiency and expanding the use of biofuels in road transport. There have been growing interest in new policy supports and fiscal incentives for electric vehicles (EVs) plus advanced biofuels for aviation and maritime transport sectors. Strong policy supports have enabled the clean renewable transport sector to grow plus to weather some of the difficulties posed by the international oil prices fluctuations (REN21, 2018).

Globally the use of biofuels for road transport has been attracting policy supports from various countries across the world. Many governments have enacted biofuel blending mandates together with various financial incentives for biofuel blending programs. These represented the most popular forms of policy and financial support for renewable bio-energy applications in the road transport sector. A good example is that new biofuel blending mandates have been enacted in Argentina, India, Malaysia, Panama and Zimbabwe. The United States have also released new biofuel blending mandates under its Renewable Fuel Standard (RFS). There have also been new policy supports for the use of advanced biofuels in the transport sector. A good example is the advanced biofuels mandates in Denmark.

Biofuels for road transport have been the priority area for government policy supports in some key countries. Policy supports for biofuel in the aviation and maritime sectors have been making slower progress. There are ongoing concerns about the sustainability of the first-generation biofuels and debates of food versus biofuel have continued especially after the recent soya oil price rises. These ongoing discussions on biofuel sustainability concerns have so far not stopped new biofuel support policies being enacted by various countries in recent years.

In Europe, the new EU package of clean energy and emissions reduction goals has provided guidance on future biofuel uses. Specifically, the EU plan called for a gradual reduction in the share of food-based biofuels in transport fuel. They have proposed reductions from 7 per cent of transport fuel consumption in 2021 to 3.8 per cent in 2030. They have also planned to increase the share of low emissions fuels, including renewable electricity and advanced biofuels, from 1.5 per cent in 2021 to 6.8 per cent in 2030.

In North America, Canada has issued a set of guiding principles for sustainable biofuels. Canada has also announced its intention to adopt a national clean fuels standard. This would build on the sub-national biofuel blend mandates already in place in five of Canada's ten provinces.

The USA has also been active with its biofuel blending policy support, which has helped to make it one of the top biofuel producers. The USA has released its 2017 blending mandates under its Renewable Fuel Standard RFS. It has planned for the blending of 73 billion litres (19.3 billion gallons) of renewable fuels. These would include 16.2 billion litres (4.3 billion gallons) of advanced biofuels and 1.2 billion litres (311 million gallons) of cellulosic biofuels. The USA has also established a mandate for blending 7.9 billion litres (2.1 billion gallons) of biomass-based diesel in 2018. At the US subnational level, Minnesota's B10 mandate is scheduled to be increased to B18. It was upheld in court after being challenged by multiple fossil fuel industry associations as being incompatible with the federal Renewable Fuel Standard. At the US state level, Hawaii has introduced a tax credit for biofuel producers, and Iowa has extended tax credits for both biodiesel and ethanol through to 2025.

In Latin America, Mexico has mandated the blending and sale of E5.8 outside of the three metropolitan areas of Guadalajara, Mexico City and Monterrey, where ethanol blending was initially piloted. Argentina has enacted a B10 and E10 mandate plus announced plans for an E26 mandate to be enacted. Panama's ethanol mandate increased to E10.

In Asia, Malaysia increased its B7 mandate to B10 and Indonesia increased its B5 mandate to B20. India also established goals of E22.5 and B15 through a new policy that promoted the use of non-conventional biofuel feedstocks. These included biofuel and biodiesel produced from bamboo, rice straw, wheat straw and cotton straw, plus ethanol from molasses. Vietnam has also established an E5 mandate. In Africa, Zimbabwe returned its blend mandate to E15 after a temporary reduction to E5 due to a lack of supply.

New financial incentives have also been introduced by various countries to promote biofuel production and consumption, bio-refinery and Research and Development (R&D) into new technologies. Two good country examples are Argentina, which has extended tax exemptions for biodiesel production and Sweden, which has introduced tax cuts on both ethanol and biodiesel.

Many countries have also introduced strong policy support and fiscal incentives for EVs. The global sales of EVs have been rising and have reached over 775,000 units globally. The EV sales currently represent only around 1 per cent of the global passenger car sales. More than 2 million passenger EVs have already been sold and are travelling on the world's roads. Looking ahead, more EVs growths are expected in the coming years. EV policy support will be discussed in more detail in the EV section below.

Global municipal green transport policy developments

Municipal policy makers from leading cities across the world have been playing increasingly important roles in promoting the use of renewable energy in their cities. More and more policy makers at the local city and municipal level have been setting new targets and enacting policies to promote clean renewables energy in their cities and towns. The growing urban population and rising urbanisation have resulted in rising energy demands for cities. The total energy consumption by cities as a share of global energy consumption has also been rising. A good example is that in 2014, cities globally accounted for 65 per cent of the global energy demands. This is an increase of over 40 per cent from the 1990 levels. Cities globally had, in 1990, consumed approximately 45 per cent of global energy demands.

All the key cities around the world have unique sets of resources and varying patterns of energy consumption. These have created unique challenges and opportunities for policy makers in different cities. A good example is that cities such as New York, London and Seoul have been using much of their energy in the building and transport sectors. On the other hand, big cities including Shanghai and Kolkata have large industrial sectors which have been accounting for the majority of their energy uses.

It is encouraging that the number of cities globally that have committed to transition to 100 per cent renewable energy have continued to grow. Some cities, such as Burlington and Vermont in the USA plus more than 100 communities in Japan have already achieved their 100 per cent renewable energy goals. Other cities have set goals of reaching 100 per cent renewables by a specific year. For example, the Australian Capital Territory have set a goal of 100 per cent renewable energy by 2020. Los Angeles, the second largest US city, has directed its municipal utility to determine how to move to 100 per cent renewable electricity, although no specific target has been established yet. Some other large cities have set less ambitious but still significant renewable and green transport targets. A good example is Calgary in Canada, which has pledged to power all government operations, including their large transport fleets, on clean renewable energy by 2025 (REN21, 2018).

Municipal policy makers have continued to make use of their purchasing and regulatory authorities to spur renewable deployments and green transportation within their jurisdictions. Government purchasing authorities have the power to change their public transportation fleets to clean fuel or EVs, plus installing solar panels on municipal buildings. A good green transport example is the London Borough of Greenwich council which has recently launched a new electric car club for all their local residents and business users. The council has teamed up with the borough's regular car club provider to trial the e-car scheme in the Low Emission Neighbourhood (LEN) of Greenwich and the Greenwich Peninsula. Electric cars will be available for hiring from fixed EV charging bays around the LEN area (Greenwich Council, 2019).

In the European green transport sector, Oslo has pledged to power its public bus fleet with renewable energy by 2020 as part of the city's green climate budget. Reykjavik has set a goal to fuel all their vehicles, both public and private, in the city with renewable energy by 2025.

In the green aviation sector, Seattle's publicly operated Tacoma Airport has become the first airport in the world to provide airport-wide access to bio-jet fuel supplies. Sacramento County in California has begun fuelling its liquefied natural gas trucks with biogas.

Municipal and city governments around the world have also been collaborating on green transport, renewable energy and climate mitigation goals. A good example is the C40 Cities initiative, which brought together leaders of 90 of the world's largest cities to discuss and launch joint pathways for leading cities around the world to meet the goals of the Paris Climate Agreement (C40 City & Arup, 2018).

Global transport innovations and disruptive trends

Looking ahead, experts have predicted that major disruptive innovation developments would likely take place which would help to transform the global transport and auto industry. These would likely revolutionise the way that industry players would respond to changing consumer behaviours, develop partnerships, and drive transformational changes. These include digitisation, increasing automation, and new business models which have already revolutionised many industries, including automotive and EV industries.

These forces have given rise to four key disruptive technology-driven megatrends in the automotive sector, which include diverse mobility, autonomous driving, electrification and connectivity (McKinsey, 2017).

Looking ahead, experts have predicted that different diverse mobility innovations, including shared mobility, connectivity services, feature upgrades plus new business models could drive major expansions globally. Experts have predicted that these could increase the automotive revenue pools globally by about 30 per cent, which could add up to $1.5 trillion globally by 2030. These should also increase on-demand mobility services and data-driven auto services.

Connectivity and autonomous technology innovations would increasingly allow the EVs to become a platform for drivers and passengers to use during transit to consume novel forms of media and services plus allow the freed-up time for other personal activities. The increasing speed of innovations, especially in software-based systems, will require cars of the future to be upgradable. As shared mobility solutions with shorter life cycles become more common, consumers will be constantly aware of technological advances, which will further increase demands for upgradability in privately used cars as well.

Overall global car sales are likely to continue to grow. However, the annual auto growth rates are expected to be lowered from the high 3.6 percent over the last 5 years to around 2 per cent by 2030. These drops will be largely driven by macroeconomic factors plus the rise of new shared mobility services such as car sharing, car clubs and e-hailing. New mobility services may result in a decline of private-vehicle ownership and sales. These declines are likely to be offset by increased sales in shared vehicles, which would need to be replaced more often due to higher utilisation, with more wear and tear.

Different auto markets around the world will also be developing different models. Their growth will be driven by their corporate strategy, macroeconomic developments and consumer preferences. The rise of the global consumer middle class, especially in the faster growing emerging economies, will be one of the main consumer driving forces for auto growth. With many developed economies' growth rates slowing down, global growths will continue to be driven mainly by the developing emerging economies.

Global consumer mobility behaviour has been changing significantly. Individual consumers have increasingly using multiple modes of transportation to complete their journeys for business or pleasure. In addition, goods and services have increasingly been delivered to consumers rather than being fetched by consumers. A good example is the increased use of online ordering, together with the quick and efficient delivery of purchased parcels to customer homes by online companies, such as Amazon.

The traditional business model of car sales has also been changing. The new range of diverse, on-demand mobility solutions, especially in dense urban environments, could discourage private-car ownership and promote shared ownerships. Looking ahead, this could lead to up to one out of ten cars being sold in 2030 potentially being a shared vehicle. On this trajectory, one out of three new cars sold could potentially be a shared vehicle by as soon as 2050.

Consumers today also use their cars as all-purpose vehicles, whether they are commuting alone to work or taking the whole family to the beach. In the future, they may want the flexibility to choose the best solution for a specific purpose, on demand and via their smartphones. There are early signs that the importance of private-car ownership has been declining in different car markets globally. A good example is in the USA, where the share of young people (16 to 24 years) who hold a driver's license has dropped from 76 per cent in 2000 to 71 per cent in 2013. On the other hand, there has been over 30 per cent annual growth in car-sharing memberships in North America and Germany over the last five years.

The new consumer habit of using tailored solutions for different driving purposes might also lead to new specialised vehicles being designed for very specific needs. A good example is the growth in the market for a car specifically built for e-hailing services. This would require a new electric car designed for high utilisation, robustness, long mileage and passenger comfort.

On market segments, cities would likely replace country or region as the most relevant segmentation dimension that determines mobility behaviour plus the speed and scope of the automotive revolution. It would be important for auto companies to acquire a more granular view of their mobility markets. Specifically, it is necessary to segment these markets by city types based on their population density, economic development, and prosperity. Across these segments, consumer preferences, policy and regulation, as well as the availability and price of new business models will also be important.

In many megacities, such as London, car ownership has been becoming more of a financial burden for many, due to new congestion fees, lack of parking, traffic jams, etc. By contrast, in rural areas, such as Iowa in the USA, private-car usage will continue to be the preferred means of transport by far. The type of city will thus become a key indicator for mobility behaviour, replacing the traditional regional perspective on the mobility market. Looking ahead to 2030, the car market in New York will likely have much more in common with the car market in Shanghai than with that of Kansas.

Autonomous vehicles are likely to grow significantly in future once the technological and regulatory issues have been resolved. Experts have predicted that up to 15 per cent of new cars sold in 2030 could be fully autonomous. Fully autonomous vehicles are unlikely to be commercially available before 2020. Meanwhile, advanced driver-assistance systems (ADAS) will play a crucial role in preparing regulators, consumers, and corporations for the medium-term reality of autonomous car developments. The market introduction of ADAS has shown that the primary challenges impeding faster market penetration have included pricing, consumer understanding, plus safety and security concerns. New high tech players and start-ups will likely play important roles in the future developments of autonomous vehicles. Regulation and consumer acceptance may represent additional hurdles for autonomous vehicles. However, once these challenges have been addressed, autonomous vehicles could offer tremendous value for consumers. A good example is the ability to work while commuting, plus the convenience of using social media or watching movies while traveling.

EVs are likely to continue their growth, as well as become more viable and competitive in future. However, the speed of their adoption will vary strongly in different countries and cities globally. Stricter emission regulations, lower battery costs, expanding EV charging infrastructures, and increasing consumer acceptance will create further momentum for EV market penetrations. Different types of EV, including hybrid, plug-in, battery electric, and fuel cell, will also have different growth rates in the coming years.

Looking ahead to 2030, the future share of electrified vehicles could range from 10 per cent to 50 per cent of new auto vehicle sales, for different markets around the world. Adoption rates are likely to be highest in developed dense cities with strict emission regulations and strong consumer incentives. Various incentives could include tax breaks, special parking spaces, plus congestion charge exemptions and discounted electricity pricing. EV market and sales penetrations are likely to be slower in small towns and rural areas with lower levels of charging infrastructure and higher dependency on driving ranges. Continuous improvements in battery technology and costs will likely make these local differences less pronounced. Looking ahead, EVs are expected to take more market share away from conventional ICE vehicles. With battery costs potentially decreasing to $150 to $200 per kilowatt-hour over the next decade, electrified vehicles should achieve cost competitiveness with conventional ICE vehicles. This will create the most significant catalyst for higher EV market penetration. At the same time, it is important to note a large portion of the electrified vehicles market segment could be hybrid electric vehicles, such as the Toyota Prius. This will mean that even beyond 2030–2050, the internal combustion engines would still be relevant globally.

Within a more complex and diversified mobility industry landscape, incumbent players will be forced to compete simultaneously on multiple fronts and cooperate with competitors. Whilst other industries, such as telecommunications, smartphones and TV, have already been undergoing disruptive transformations, the automotive industry has seen few major disruptions so far. For example, only a few new auto players have appeared on the list of the top 15 automotive original-equipment manufacturers (OEMs) in the last 15 years compared with the list of top 15 new players in the mobile handset industry, which has seen many more changes globally.

A paradigm shift to mobility as a service, along with new auto entrants, will inevitably force traditional car manufacturers to compete on multiple fronts. The emergence of different new EV players has increased the complexity of the competitive landscape. These include mobility providers such as Uber, tech giants such as Apple and Google, plus speciality original equipment manufacturers (OEMs) such as Tesla. Traditional automotive players would be facing more pressure to reduce costs, improve fuel efficiency, reduce emissions plus become more capital-efficient. These changes will likely lead to shifting market positions in the evolving automotive, EV and mobility industries. They are also likely to lead to more consolidation or newer forms of partnerships among incumbent players.

The growing and diverging auto markets will open opportunities for new players. However, these new market players will initially focus on a few selected steps along the value chain. They will probably target only specific, economically attractive market segments before expanding further into other selective areas. Good examples include Tesla, Google and Apple, which have all been showing interest in the growing EV and AEV markets. Many more new players are likely to enter the market, especially other cash-rich high-tech companies and start-ups. Similarly, some Chinese car manufacturers, with impressive sales growths recently, might leverage the ongoing disruptions to play more important roles globally.

Automotive companies have to develop good strategies and make important strategic moves now to accommodate the inevitable disruptions. They would need to prepare for uncertainty and shift to a continuous process of anticipating new market trends, revising their traditional business model and exploring new mobility business models. These will require more sophisticated scenario planning, enterprise risk management plus organisational agility to identify new attractive business opportunities quickly.

Auto players will also need to leverage potential partnerships for future growth. The auto industry is transforming fast from competition among peers toward new partnerships and alliances. To succeed, automotive manufacturers, suppliers and service providers would need to form new alliances or participate in specific ecosystems. These would include new infrastructure for autonomous and electrified vehicles.

With innovation and product values increasingly defined by digital systems and software, OEMs would need to align their skills and processes to address the new digital challenges including digital transformation, software-enabled consumer value definition, cybersecurity, data privacy and continuous product updates.

Car manufacturers must further differentiate their products and services as well as change their value propositions from traditional car sales and maintenance to integrated mobility services. This would put them in a stronger position to gain future shares of the growing global automotive market plus its revenue and profit pools. They would also need to embrace new business models such as online sales and mobility services, plus cross-fertilising the opportunities between the core automotive-business and new mobility-business models.

5 Climate change electric vehicle growth management

船到桥头自然直
Chuán Dào Qiáo Tóu Zì Rán Zhí
As the ship reaches the bridge, it has to align with the bridge to berth.
Cross the bridge when one comes to it.

Executive overview

Globally the transport sector and ICE vehicles have contributed significantly to GHG emissions, which have led to global warming and climate changes. Rising climate change concerns and new green transport policies from different countries have promoted the growth of EVs globally. Transport experts have predicted that EV sales could rise to 100 million by 2035 and then further rise to over 200 million EVs globally by 2040. The new EV growths could cover any road-, rail-, sea- and air-based transport vehicles that would use electric drives. The electric charging of EVs will also promote integration with clean renewable sources from both city grids or home charging points. Details of EV growth management and technological innovations will be discussed further in this chapter.

Global electric vehicles (EV) growth and developments

The transport sector and ICE powered motor vehicles globally have been major contributors to GHG emissions and global warming. Climate change concerns and the green transport drives globally have promoted the growth of EVs globally. Transport experts have forecast that increasing numbers of EVs will be joining the global fleet in future. Many of these forecasts have predicted that EV globally will rise to 100 million by 2035 and then further rise to over 200 million EVs globally by 2040.

Electric vehicles encompass any road-, rail-, sea- and air-based transport vehicles that would use electric drives. The electric charging of EVs can come from an external electrical source such as from city grids or home charging points. Some EV technologies have been hybridised with fossil fuel engines, such as PHEVs. A plug-in hybrid electric vehicle (PHEV) is a hybrid electric vehicle whose battery can be recharged by plugging it into an external source of electric power, as well by its on-board engine and generator. Good examples of PHEV include the Audi A3 E-Tron, Audi Q7 E-Tron plug-in hybrid or BMW 330e. Looking ahead, a new third variant of EV would be the use of fuel cells to convert hydrogen into electricity. Smart EV charging systems could also help to integrate growing quantities of variable renewable energy for EV charging applications (REN21, 2018).

Electrification of the transport sector globally has been expanding. These expansions have enabled the greater integration of clean renewable energy in the form of green electricity for trains, light rail, trams as well as two- and four-wheeled EVs. Political interest in electric mobility has increased following the Paris Agreement, which sparked off a broader debate on accelerating electrification of the transport sector. (REN21, 2018).

Global deployment of EVs for road transport, and particularly passenger vehicles, has grown rapidly in recent years. However the EV passenger car market (including PHEVs) is currently only accounting for around 1 per cent of global passenger car sales. The top 5 countries for passenger EV deployment have been China, the United States, Japan, Norway and the Netherlands. Together they have accounted for over 70 per cent of global EV sales. China and the USA have been the market leaders in EV unit sales. Norway has been the market leader in terms of EV market penetration.

China's EV market has seen dramatic growth in recent years. EV sales nationally have increased from about 11,600 vehicles in 2012 to more than 350,000 in 2016. China has overtaken the USA in 2016 to become the country with the largest number of passenger EVs on its roads.

In most countries, even those with strong incentives, EVs have continued to represent a small share of the passenger vehicle sales currently. Norway is the only market in which EV market shares have reached a mass market stage and it has the highest EV market penetrations globally. These high market penetrations have been driven by a set of strong government incentives, which include EV exemption from sales and registration taxes, as well as the construction of extensive national EV charging infrastructures. EV sales have represented 29 per cent of new passenger vehicle registrations in Norway.

Although electrically driven passenger cars have experienced the most rapid market growths in recent years, electrical drives have also been applied in other transport modes including trains, trams, buses, two- and three-wheeled vehicles. A good example is that in Europe, some 5,500 electric buses have been operating on the roads. Around 90 per cent of these electric buses were connected via overhead electric wire networks linked to the city grids.

EV manufacturers and electric utilities have been experimenting with 'smart' charging and vehicle to grid technologies. These would enable EVs to be charged by variable clean renewable energy with grid energy storage. A good example is the Netherlands, which has become an international leader in the use of variable renewables for EV charging via new advanced smart-charging systems.

There are significant challenges for EV scale-ups to meet future market expectations. Some of the challenges include limited EV drive ranges, limited availability of charging infrastructure, and a lack of uniform charging standards. There are also three different plug types for the rapid charging of EVs. Firstly the CHAdeMO network, which works only with Asian-made vehicles. Secondly, the SAE Combo plug, which fits in German and some US-made vehicles. Thirdly Tesla's Supercharger network, which only fits Tesla electric vehicles. These three potential standards are currently all competing in the marketplace and they should be harmonised if possible.

The growth of EVs will also require a dramatic scale-up in the lithium-ion battery supply chain. Lithium-ion battery manufacturing capacity today is around 131 GWh per year. Based on plants announced and under construction, these are likely to increase by about 4 times, to over 400 GWh by 2021 with some 73 per cent of the global capacity concentrated in China.

Battery improvements would be critical for promoting future EV sales. The price of lithium-ion battery packs, which was $1,000 per kilowatt-hour in 2010, has now been reduced by three-fold to below $300. Experts have forecast that battery pack prices could be further reduced to less than $100 by 2026. These lower prices should then further promote EV sales globally (BNEF, 2018).

Electric vehicle enabling policies and fiscal supports overviews

There are various key drivers for new policies to support EV developments in different countries globally. These include enhancing energy security, reducing transport-related carbon emissions and increasing the opportunities for sustainable economic growth. Most of the EV support policies enacted by different governments globally have been aiming to reduce air pollution and to lower GHG emissions. These will help to improve air quality in major cities around the world and thereby also improve public health. The future targets for EV have been articulated by some countries and provinces in terms of zero-emission vehicles (ZEV), which are largely synonymous with EVs. These ZEVs will include both EV and PHEVs (REN21, 2018).

A new International Zero Emission Vehicle (ZEV) Alliance has been established by several key European countries plus leading North American municipalities, including US states and Canadian provinces. The ZEV Alliance announced in late 2015 a common goal to achieve zero emissions for all new cars by 2050 (ZEV Alliance, 2015).

Various leading European countries have set strict ambitious targets for EV build-ups. A good example is the Netherlands, which had set targets in 2016 for 10 per cent of new cars to be EVs by 2020, 50 per cent by 2025 and 100 per cent by 2035. Another good North European example is Norway, which committed to all new passenger cars, city buses and light vans to be EV or ZEVs by 2025. Norway has managed to have 50,000 EV and ZEVs on its roads three years ahead of their original plans. In the United Kingdom, all new cars and vans must be EV or ZEVs by 2040. In addition, the UK government has set the new goal that nearly all cars and vans running on UK roads should be EV or ZEVs by 2050.

In Asia, both China and India have announced aggressive plans for EV and ZEVs. India has issued its new National Electric Mobility Mission which set new targets to have 6 million EVs, including hybrid PHEVs, on their roads by 2020.

China issued their new 'Technical Roadmap for Energy Saving Vehicles' in October 2016. The China roadmap set a target for 7 per cent EV sales by 2020 and 40 per cent EV sales by 2030. China has also issued a new target for the development of EV charging infrastructure nationally. They are aiming to build 12,000 new charging stations across China to serve the 5 million EVs on their roads by 2020 (PRC, 2016).

In the United States, some states including California, have enacted EV policies which had stipulated that EV and ZEVs would make up around 15 per cent of new car sales by 2025. California state government has also stipulated that the renewable energy share of hydrogen for vehicles should increase to 33 per cent by 2022 as part of their policy of support for new hydrogen vehicles and low carbon economy developments.

Fiscal incentives and supports have been used by different central and municipal governments to promote EV sales in their countries and cities. In Europe, Germany launched a support scheme for EVs in 2016. The fiscal supports have included EV purchase grants and funding to expand recharging infrastructure. The Austrian government has also offered a purchase premium for EVs charged with 100 per cent renewable electricity from 2017 onwards.

In Asia, Japan has offered subsidies for the purchase of low-emissions vehicles, including EVs. China has also extended their national New Energy Car Program to 2020. The program included fiscal incentives plus preferential registration processes in local government agencies. Experts estimated that the China fiscal supports have included some USD 4.5 billion in subsidies for the purchase of EVs in China. Looking ahead, China is planning to gradually phase out their New Energy Car programme by 2021.

Some leading cities around the world have also been developing new strict zero-emission transport targets and policies. A good example is Amsterdam in the Netherlands, which has set new municipal government targets committing Amsterdam to become a zero-emission city by 2025. Amsterdam also began in 2018 to replace all their 200 public transit buses with electric buses. In addition, Amsterdam has stipulated that they plan to replace 4,000 taxis with EVs and ZEVs under the Clean Taxis for Amsterdam Covenant.

In China, Taiyuan City has become the country's first city to replace its entire taxi fleet with EVs. The Taiyuan City government has also funded and built a new network of 1,800 charging stations for EVs across its city. In addition, at least 14 leading Chinese cities, including Beijing and Shanghai, have offered subsidies to developers to encourage the building of new EV charging stations in their cities. In addition, Beijing municipal government have exempted EVs and ZEVs from driving restrictions that they have imposed on internal combustion vehicles. Their current restrictions on petrol and diesel ICE vehicles have stipulated that these vehicles will not be permitted to be driven one day per week. There will also be restrictions on new licence plate registrations for fossil ICE cars, with licence allocation by lottery.

Key electric vehicle companies developments overviews

The rapid emergence and growth of electric drives as a serious alternative to the traditional internal combustion engines ICE has opened many new opportunities for new market entrants to the automotive market. Good examples include Tesla and BYD, who have both quickly became global market leaders in EV manufacturing. Tesla was founded as an EV company in 2003 in the USA, whilst BYD was started as a battery manufacturer in China. Both Tesla and BYD have become new global market leaders for passenger EVs. BYD from China has been selling some 100,000 vehicles annually and has achieved an impressive 13 per cent market share in China and globally. BYD was started as a battery manufacturer in 1995 and has been a relative newcomer in the automotive industry (REN21, 2018).

Renault-Nissan (France-Japan) has become the global market leader in terms of cumulative sales of EVs across different countries. It has sold over 350,000 EVs globally. The other leading global EV players include Tesla from the USA and BMW from Germany.

Several long-established traditional ICE vehicle manufacturers have also been realigning their strategies and plans so as to increase the share of EVs in their future sales. A good example is the Volkswagen Group VW from Germany, which has recently announced plans to bring more than 30 pure-electric models to the car market in the next few years. VW has set new sales targets of selling 2 to 3 million EVs annually by 2025 which will be equivalent to 20–25 per cent of its total projected global sales of all vehicles. As part of their new strategy covering both ICE, EV and ZEVs, VW has announced new R&D plans to research and develop new battery technology as a new core competency. VW has also expressed interest in building its own battery factory to supply battery for their growing EV productions. Another good example is Daimler AG in Germany, which announced in 2016 that it would invest USD 10.5 billion or EUR 10 billion in EVs manufacturing. Daimler is planning to design and manufacture 10 new different EV models by 2022.

A new emerging global megatrend is that some successful large high tech companies have been showing interest in becoming involved in the EVs business. A good example is Apple in the USA, which has started to invest in EVs. Apple has been spending more on R&D in recent years on

EV vehicles and related services than it has done on several other Apple products combined. Google has also been interested in EV applications, especially on autonomous electric vehicles (AEVs). They have set-up a special business unit which has been actively researching and developing AEV driving applications.

In addition, several other global consumer electronic companies have announced their interest in entering the global EV market. There have been a lot of activities particularly in China. Some 200 mostly small medium enterprises SMEs in China have been reported to be developing and marketing EVs. They aspire to follow the successful footsteps of BYD.

There are some serious development hurdles for EV and ZEVs. These include their limited driving ranges plus their battery capacity and life expectancy. Many manufacturers have been working on advancing battery technologies so as to increase the driving ranges of new EVs. A good example is the two mid-priced battery-EV models from Renault-Nissan and General Motors from the USA which have recently entered the EV market with driving ranges of more than 300 kilometres each. Several other EV companies have also announced plans to launch new EV vehicles with equal or greater driving ranges in the coming years.

Many EV manufacturers have announced plans for new EV models with driving ranges of up to 500 to 600 km. A good example is Daimler AG, which has announced its intention to introduce a new EQ battery EV It is hoped it will have a driving range of up to 500 kilometres and they are planning to launch it by the end 2020. Volkswagen Group is also planning to introduce a new concept e-Golf model with a projected range of up to 600 kilometres. VW is planning to introduce the new model to the market in 2020 at a similar cost to its diesel-based equivalent.

On EV charging systems, various electric vehicle manufacturers and energy industries have been actively developing new EV charging stations as well as addressing the shortage of charging stations in different countries. The European Union countries have worked together to expand the EV charging infrastructure from 30,000 stations in 2014 to over 100,000 stations, which also includes 10,000 fast charging stations.

Several leading auto manufacturers, including BMW Group from Germany, Daimler AG, Ford Motor Company from the USA and the Volkswagen Group have announced that they will be jointly forming an important new joint venture on EV charging systems. Their new joint venture began in late 2016 and has plans to deploy, starting in 2017, a new network of high-powered 350 kW charging stations i n Europe. These should enable long-range travel and charging for EVs across different EU countries. These new rapid charging stations will have capacities that are more than double the 2016 capability of Tesla Superchargers. These rapid charging stations should allow EVs with a range of 400 kilometres to reach a full charge in only 12 minutes. These much faster charging times should be attractive to EV owners and should help to promote more customer acceptance of EVs. Some companies have even advertised that EV owners would be able to enjoy a cup of coffee at their station café whilst their EVs are being rapidly charged at their new charging stations.

In the USA, Nissan and BMW have also announced plans to install fast-charging stations across the country that would be equipped to work with both CHAdeMO and SAE Combo connectors. US electric utilities have also joined the effort to expand charging infrastructure, but some have been blocked by regulators over concerns about who should pay for what.

Reducing battery costs and recycling are important strategic drivers for EV market developments and growths in future. For EV competitiveness in general, the mega-trends towards longer battery lifetimes and higher energy storage densities plus green recycling are the critical future development requirements.

On battery costs, General Motors has announced that its battery cell cost for the Chevrolet Bolt will be lowered to USD 145 per kWh. This would be a very competitive battery cell cost when compared to the normal aver-aged battery costs of USD 150 to 250 per kWh. The battery sizes for small and mid-sized battery EVs have ranged from 30 kWh up to a maximum of 60 kWh. Tesla has started to offer bigger battery capacity of up to 100 kWh battery capacity. EV manufacturers have been taking advantage of lower battery prices to increase the driving range of their EVs.

In addition to passenger EVs, many auto makers have also been working on electric transport developments for public transit, railway and freight transport. A good example is that Siemens from Germany announced that they have made significant advances with their long-distance pure-electric trucks which could transform the European freight transport sector. In addition, different innovative EV companies in California, Singapore and Switzerland announced that they have been actively researching and explor-ing the potential of autonomous electric buses.

Clean renewable energies are also increasingly being used in different countries globally to charge the public transit systems. A good example is that Chile announced in 2016 that Santiago's subway system, which is the second largest in Latin America after Mexico City's, will be powered mostly by solar PV and wind energy as part of their national renewable energy drives.

In Africa, Uganda launched Africa's first solar-powered bus which will use an electric battery charged with solar power to extend the range of the buses. An Australian company announced plans to launch a solar-powered jeepney for use in the Philippines. In Bangladesh, an inexpensive solar-powered three-wheeled ambulance will be used to provide service to rural areas of Bangladesh.

In the aviation sector, there has been significant research into electric airplanes. A good example is the solar powered aircraft, called the Solar Impulse 2, which successfully completed an around-the-world flight after a 16-month voyage recently.

The research on exploring different optimal methods to integrate renewable energy into charging stations for electric cars have also been growing. Many research projects are still at the pilot or demonstration stages and the integration has remained relatively small-scale. A good example is the installation of what is reportedly the world's first solar-controlled, bi-directional charging station for EVs which was completed in Utrecht, in the Netherlands. This was undertaken as part of the Netherlands national Living Lab programme.

There have been other innovative offerings in EV service provisions emerging globally. Some EV service providers, including car sharing companies in the UK and the Netherlands, have begun offering new provisions for their customers to buy renewable electricity to charge their EVs.

In addition, an increasing number of EV companies have been working to integrate renewable energy technologies directly into vehicles. A good example is the Hanergy Holding Group (China), which has introduced four concept EVs which will use solar power to extend their driving ranges. Hanergy announced they have plans to produce these solar EV vehicles commercially within three years.

EV and battery technology developments and implications

Some of the key success factors for future EV market developments and growths will include battery improvements and cost reductions. Future innovations for longer battery lifetimes and higher energy storage densities will be very important to improve the driving ranges and lifetimes of future new EVs. In addition, the development of suitable new battery systems for green recycling will be important to meet the new environmental and recycling requirements that have been introduced by different countries globally in line with their Paris Agreement commitments.

Battery experts have generally agreed that the prices for battery and energy storage systems should be falling further in coming years with various innovations and technical break-throughs. There have been some large disparities and disagreements over how far and how quickly these battery cost reductions will proceed. These are important since significant battery prices reductions would have wide-ranging effects across various industrial sectors, for customers and key markets globally. In particular, cheaper batteries could enable the faster adoption of electrified vehicles globally. This should lead to major changes in not just the transport sector but also the power and petroleum sectors globally.

Latest economic research has suggested that the price of lithium-ion batteries could fall dramatically during 2020–2025 with various innovations. Recent expert analysis has indicated that the price of a complete automotive lithium-ion battery pack could fall from current costs of $500 to $600 per kilowatt hour (kWh) to about $200 per kWh by 2020–2021 and to about $160 per kWh by 2025. This should help to create more favorable conditions for the wider adoption of electrified vehicles in various key markets globally. For example, in the USA, with gasoline prices at or above $3.50 a gallon, automakers who could acquire batteries at prices below $250 per kWh would then be able to offer their electrified vehicles competitively, on a total-cost-of-ownership basis, as compared to traditional vehicles powered by advanced internal-combustion engines ICE (*McKinsey Quarterly*, 2018).

The pace of EV adoption globally will also hinge on a range of strategic factors in addition to battery prices. These would include macroeconomic and regulatory conditions, the performance and reliability of the vehicles, battery recycling and customer preferences. The rates at which automakers could achieve future battery improvements and prices reductions would also vary depending on the different manufacturing, investment and power train–portfolio strategies adopted by each of these auto companies. Looking ahead, the development of suitable new battery systems for green recycling would also be very important to meet the new environmental and recycling requirements that are being introduced by different countries globally in line with latest environmental requirements and their Paris Agreement commitments.

EV manufacturing research has shown that various EV manufacturing factors would be key in determining EV costs and profitability. The key manufacturing factors would include manufacturing at scale and productivity improvements plus EV design improvements. These could represent about one-third of the potential price reductions of EVs through to 2025. EV manufacturing cost savings could come largely from improving manufacturing processes, standardising equipment, and spreading fixed costs over higher unit volumes. Future new EV manufacturing plants could be significantly more productive than those in operation before 2010, as they would be able to adopt the latest manufacturing processes and technologies quickly. A good example is the new major EV manufacturing plant that Tesla is building in China.

The reductions in EV materials and components prices could generate up to about 25 per cent of future EV cost reductions. Global competitive pressure should drive component suppliers to reduce their costs dramatically. These could include increasing manufacturing productivity and moving operations to locations globally where costs are optimal. New EV materials, particularly those which could combine light weighting with high performance, will become more important. The new lighter stronger materials, such as thermoplastic, polycarbonates or auto copolymers, could help to reduce the overall weight of future EVs whilst meeting the demanding external appearance and impact resistances requirements.

Battery capacity-boosting technology innovations will be important for improving EV battery capacity and life. Technical advances in cathodes, anodes, and electrolytes could increase battery capacities by 80 to 110 per cent by 2020 to 2025. These improvements could generate over 40 percent of future battery cost reductions. A good example is that new battery cathodes that incorporate layered structures would help to eliminate dead zones, which could improve battery cell capacity by 40 percent. Manufacturers have also been developing high-capacity silicon anodes which could increase cell capacity by 30 per cent over the graphite anodes that are currently being used today. Researchers are also developing new cathode–electrolyte pairs which could increase cell voltage to 4.2 volts, from 3.6 volts, by 2025. These should increase cell capacities by some 15 to 20 percent plus over current present-day standards.

EV automakers will have to carefully balance their projections of the pace and trajectory of improving battery and energy storage prices against how other power train technologies and fossil fuel prices will develop. A good example is that future scenarios which would project a relatively quick decline in battery prices plus flat or rising petroleum prices would favor battery electric vehicle (BEV) development. On the other hand, those scenarios anticipating slower declines in battery prices, as well as increases in petroleum prices, would more favor plug-in hybrid-electric vehicles (PHEV). EV manufacturers should also develop suitable enterprise risk management systems and risk mitigation strategies to manage these potential development risks. These should then allow them to hedge their bets in light of the long product development cycles and choose the right investments in a range of potential new technologies.

It is important to note that many innovations that will enable battery improvements and price reductions for automotive lithium-ion batteries are likely to be realised first in other faster moving consumer sectors, such as consumer electronics and smart-phones. The global demands for cheaper and better-performing batteries in these consumer sectors have been intense and are critical for their continual growth and successes. Other industrial sectors could also face big disruptions as well from battery and energy storage improvements. The emergence of cheaper, improved battery and energy storage should promote faster growth of variable renewable power and lower the profitability of the traditional capital-intensive fossil fuels based assets in the electricity generation and petroleum sectors. A good example is that power companies globally could face new disruptive challenges if low-cost battery storage were to enable the wider use of distributed variable renewable power generation globally. Another disruptive example is the adoption of electrified-vehicle charging could alter the patterns of electricity demands in many key markets. In addition, the competition between EVs and advanced ICE vehicles could accelerate the reductions in demands for fossil transport fuels globally. Hence the oil companies and oil refiners globally will have to plan ahead and rethink their long term oil products strategies and portfolios.

Regional electric vehicle and battery supports case studies

Many governments globally have been setting up new schemes to support EV developments plus promote battery research and improvements. Two good examples are the UK and China. Details of their new schemes and supports will be discussed in more detail below, with relevant examples.

In UK, the UK Government Business and Energy Secretary has launched the first phase of a new government scheme, known as the Faraday Challenge. This included a £45m government funding to establish a centre for battery research which aims to make the UK a world leader in the design, development and manufacture of electric batteries. The new centre will be spearheaded by the Engineering and Physical Sciences Research Council (EPSRC) to create a new battery R&D institute to make new batteries more accessible and affordable. UK has plans for three-fifths of new cars in UK to be electric by 2030. The UK Petroleum Industry Association have reported that there are currently some 37.5 million vehicles licensed for use on roads in Great Britain. Fossil-fuelled vehicles, both petrol and diesel, are currently accounting for 99 per cent of the current UK's passenger car fleet. Hence, to achieve the large ICE vehicles to EV transition, the UK Government has launched the new Faraday Challenge to promote innovations in green energy storage and battery technologies. This is the first phase of a four-year £246m UK Government investment into new battery innovations. The longer-term UK plans include creating giant battery facilities around the National Grid to store excess variable wind and solar energy to meet peak electricity demands. In addition, new regulations will be introduced to help households with solar panels to generate and store their own electricity with new battery technology plus sell the surplus power back to the Grid when they do not need it. The UK will also reduce the costs for people and businesses who power down appliances at peak times and use electricity at cheaper times. The UK Government and UK Energy regulator Ofgem have estimated that consumers could save between £17bn and £40bn by 2050. In addition, these will also promote the growth of EVs and ZEVs plus create new green tech jobs in the UK.

The PR China Government has been actively supporting their emerging domestic electric car industries and has extended the China New Energy Vehicle Program to 2020. In June 2012 China first issued their plans to support the domestic EV manufacturers. Their plan had set an initial sales target of 500,000 new energy vehicles, including EV, by 2015 which will then rose to 5 million by 2020. In September 2013, the Central Government introduced a New Energy Car Subsidy Scheme which provided a maximum subsidy of US$9,800 towards the purchase of a new electric passenger vehicle in China and up to US$81,600 subsidy for a new electric bus in China (Wang, *Energy Markets in Emerging Economies*, 2017).

The Chinese Government's support for New Energy Vehicles and EVs has helped the growth of new EV companies in China. Some world-class EV manufacturers have emerged in China which have helped to create new jobs and exports. In addition, the new EV growth has helped to reduce China's fossil oil consumptions and improve its energy security. It has also helped to reduce air pollution and carbon emissions in China. China currently has the world's largest fleet of light duty plug-in electric vehicles. China has also overtaken both the USA and Europe in the cumulative sales of electric vehicles. China has become the world's largest plug-in electric car market, with record annual sales of more than 200,000 plug-in electric passenger cars, which represented over 34 per cent of global sales. China has also become the world's largest electric bus market, with over 173,000 plug-in electric buses. By 2020–2025, China is expected to account for more than 50 per cent of the global electric bus market. China's new energy car company, BYD Auto, has overtaken both Mitsubishi Motors and Tesla Motors to become the world's second largest plug-in electric passenger car manufacturer, after Renault-Nissan.

Looking ahead, the Chinese new energy and electric car manufacturers will have to compete with other electric car manufacturers in both the domestic market and international markets. Hence it is very important that they continue to innovate with new products and manage their cost of productions at globally competitive levels. A good example is that the Chinese government has approved Tesla to build a new world-class EV manufacturing complex in China, with the latest advanced technologies. This will further intensify competition in the EV markets in China and globally.

6 Climate change green agriculture growth management

虎父无犬子
hǔ fù wú quǎn zǐ
The sons of lions will not become like dogs.
Strong father, strong sons.

Executive overview

The agricultural sector and farm animals have been contributing to GHG emissions and global warming. The serious challenges posed by climate change to agriculture and sustainable food security will require joint action from government and the agricultural sector. It will be essential to link green agriculture with the food and human supply chain on a sustainable basis. Looking to the future, green agriculture developments will be important contributors to sustainable food and crop production whilst helping to decrease GHG emissions and reducing global warming. The key developments in green agriculture together with new innovations and key international co-operations will be discussed in this chapter, with international examples.

Climate change implications on green agriculture developments

Green agriculture is one of the most important sectors for many countries globally in terms of its potential to influence a wide range of issues that are critically related to sustainable development. These key issues include economic growth, employment, food supply security, trade flows, poverty, human health, climate change, biodiversity, etc. In addition, agriculture has close linkages with the optimal use of natural resources, especially land and water. Despite the increases in farming productivity and yields over the past few decades, the agricultural sector has been characterised by declining growth rates and productivities. These have led to decreasing share of global agricultural exports from developing countries despite heavy global farming subsidies of over U$1 billion. The growing uses of agrochemicals and chemical fertilisers have also led to serious negative impacts on human health, ecosystems, and biodiversity. Farming have generated high GHG emissions, especially from livestock and farm burnings, which have contributed to global warming and climate change impacts globally.

Climate change and global warming have had profound influences on the appropriate agro-ecological conditions required for sustainable green agriculture and secured food production in many countries globally. These serious challenges include environmental pollution, biodiversity losses, soil degradation, erosion, water scarcity, carbon foot-print and natural resources depletion, etc. As a result, various governments, farmers and rural populations have been developing their green sustainable agriculture strategy and practices in order to continue their livelihood, manage their natural resources and achieve food security on a sustainable basis (OECD, 2010).

The serious challenges posed by climate change to green agriculture and sustainable food security will require a holistic and strategic approach from government and stakeholders. These will be essential to link the various key elements of the green agriculture supply chain plus interdependencies across the whole agriculture and food systems. Looking to the future, green agriculture developments should help to improve food production whilst helping to lower GHG emissions and optimise the use of natural resources.

According to the United Nations, green agriculture should incorporate ideas and guidelines from different conceptual areas. These include key areas such as fair trade, ecological agriculture, organic or biodynamic agriculture, conservation agriculture plus agricultural innovations. Basically, green agriculture should use well-developed modern farming and sustainability concepts to improve natural agricultural techniques, dealing with things such as weed and pest management and organic fertilisers and seeds. It should also draw on agricultural technology innovations and advances to improve farming practices and sustainability. These should complement natural farming methods and new innovations such as new, synthetic fertilisers and modified seeds, etc. Green agriculture should use adaptable local farming techniques to increase farming yields and reduce waste plus support sustainable ecosystems. Green agricultural practices should also provide a higher return on land, labour and investments, etc. (UN, 2012).

The UN has highlighted five key principles for green sustainable agriculture. These include integration of livestock and crops, using post-harvest storage and processing facilities to reduce waste, making sure crop rotations are diversified, using environmentally sustainable weed and pest control practices plus the use of natural and sustainably made nutrient inputs. These should combine both sustainable environmental practices with better labour usages and application on farms globally. These should also help to reduce poverty and improve returns on farming investment and labour globally. The smooth transition to green agriculture from traditional farming practice will be critical to achieve success. UN suggested that there are six essential elements for a smooth transition. These would include diversifying crops and livestock, making water uses more sustainable, managing soil fertility, making sure agriculture storage facilities are efficient and sustainable, improving the management of animal and plant health plus introducing new environmentally friendly and labour-friendly modes of mechanisation on farms.

Green agriculture is a wide-ranging and complex approach which draws upon ideas that would need to be jointly introduced by local, national and global governments and organisations. Green agriculture can also be practiced successfully by both large and small-scale farmers. Conventional and traditional farmers can also take on these sustainable practices and methods to improve their agricultural results. Details of green agriculture will be described further in the sections below, with international examples.

Climate change agriculture sustainability concerns

Climate changes have exerted profound influences on the agro-ecological conditions for sustainable agriculture globally, especially changes in climatic and ecological dynamics. These have led to severe extreme weather events, such as hurricanes, heavy downpours, flooding and droughts, which have been very damaging for agriculture. In addition climatic and ecological changes have impacts across several interrelated societal sectors including agriculture, forestry, fisheries, health, energy, economy, employment etc. We have to take account of these complexities to develop the appropriate improvement strategies for green sustainable agriculture (OECD, 2010).

Some agricultural practices have caused serious negative environmental and sustainability consequences to date. These include the use of nitrate fertilisers and pesticides in farms globally, which has given rise to serious residue pollutions. There has been serious loss of biodiversity with agronomic practices such as monocropping. Monocropping has led to destruction of natural habitats plus over-exploitation of natural resources and biodiversity losses. There have also been serious natural resource degradations and salinification associated with artificial irrigation. The agricultural water use foot-prints has been high due to excessive water use in intensive agricultural production. There has also been serious competition with other water users, particularly drinking water and groundwater contaminations by chemicals and fertilisers. The agricultural carbon footprints have risen due to fossil fuel uses in farming and farming practices plus the production of chemicals and fertilisers from fossil feedstocks. The transportation of agricultural produces from rural communities to markets in big cities has also created traffic issues and high truck emissions.

Globally, agriculture and farming have been major contributors to climate change and emissions. These have included high CO_2 and methane emissions by livestock and farming practices. Some traditional farming techniques have generated high environment impacts and pollutions, such as the burning of fields after harvesting by farmers in many developed and developing countries. On the other hand, agriculture also has an important role in climate change mitigation strategies. A good example is biofuel production plus a range of bio-production systems with beneficial environmental consequences, such as carbon sequestration. A good example is that improvements in farm livestock management in Canada have helped to reduce the carbon footprints of beef by up to 15 per cent per pound recently. More work will be required to decrease the farming carbon footprint further globally (Agriculture and Agri-Food Canada, Climate Change and Agriculture, 2018).

The unprecedented period of agricultural intensification in most OECD countries in the last few decades has created serious negative environmental consequences. The term 'agricultural treadmill' has frequently been used to explain how the development of agriculture in developed countries have resulted in a cycle of negative environmental consequences. A serious example is farmers becoming dependent on chemical pesticides, which has led to biodiversity damages and pest resistances. This has resulted in stronger pesticides required to maintain effective pest controls, resulting in farmers being trapped on the agricultural treadmill. Agricultural intensification in the past decades has also led to land and soil degradation, salinisation of water resources, pesticide pollution of soil and water, food chain disruptions, depletion of ground water plus genetic homogeneity of agricultural products etc. All of these have raised serious concerns about the sustainability of modern intensive agriculture practices. These highlight the urgent need to transition to green sustainable agriculture.

Water scarcity and salinity have been serious environmental problems created by agricultural intensification and climate change. Water is becoming scarcer and more expensive globally. Artificial irrigation systems have been a double-edged answer to water scarcity. Common problems have included waterlogging and salinity resulting from excessive water usages and poorly designed drainage systems. The falling groundwater levels and rising pumping costs have also become serious environmental issues. Contamination of groundwater by fertilisers and chemicals has been rising globally.

Soil degradation with serious loss of nutrients, soil retrogression and erosion plus land-cover change have become serious threats to sustainable agricultural production around the world. The soil crisis has largely been brought to a head by conventional modes of agriculture intensification. Intensive chemicals and pesticide applications have contributed to the loss of essential soil nutrients. Monocropping, intensive mechanised tilling or ploughing practices plus overgrazing, land-use conversion and deforestation have all contributed to serious soil degradations globally.

Pest control has become an increasingly serious constraint on agricultural production despite dramatic advances in pest control technology. A good example is that in the US, pesticides have been the fastest growing input in agricultural production over the last half-century for pests such as pathogens, insects and weeds. Pest control in past decades has involved the application of various toxic chemicals. A good example is the pesticide Dichloro-diphenyl-trichloroethane (DDT). It was discovered in the 1930s and applied extensively as it was relatively inexpensive and effective at low application levels. Chemical companies have also introduced a series of other synthetic pesticides in the 1950s. The initial effectiveness of DDT and other synthetic organic chemicals has led to the neglect of other pest control strategies. By the early 1960s, scientific evidences showed that these synthetic chemical pesticides have caused serious environmental damages. These included direct and indirect damages on wildlife populations plus human health. Additional environmental costs involved the destruction of beneficial insects and the emergence of pesticide resistance in many target pest populations. A fundamental problem in intensive pests control is these have resulted in evolutionary selective pressure for pests that would be resistant to the chemicals and pesticides.

The rising global temperatures and increase in the atmospheric concentration of GHGs have also affected agriculture globally. Increasing global temperatures have serious implications on crop production and growth. Rising sea levels have resulted in inundation of coastal areas including paddy fields. The intrusion of saltwater into groundwater aquifers has seriously affected the supply of suitable groundwater for farming irrigation purposes.

Farming experts have defined green sustainable agriculture as agricultural practices which are capable of maintaining its productivity and usefulness to society indefinitely. These new green sustainable agriculture approaches will use farming systems that conserve resources and protect the environment plus produce crops efficiently. In addition, these green agriculture practices should be able to compete commercially and enhance the quality of life for farmers and society overall (UN, 2012).

A good example of green agriculture is the Dutch agriculture approaches. The Netherlands has one of the most intensive farming systems in the world, with high output levels supported by a considerable use of agrochemicals. As one of the smallest countries in the EU, the limited availability of agricultural land has led to rising agricultural intensification over time. The growth of the EU common market has promoted free internal trade within the EU. These have resulted in the Netherlands being in the top three agricultural exporting nations in the world.

The Dutch policy-makers have long been concerned over issues of environmental sustainability as a result of agricultural intensification. Key areas of sustainability concerns included pollution of groundwater, ammonia emissions and their impact on the acidification of soils and water, negative effects of pesticide use, biodiversity and landscape issues, etc. The Netherlands has been one of the first countries globally to make system-wide farming changes to transition to green agriculture so as to address these serious concerns in the early 1990s. The Netherlands has one of the longest histories of policy development to restrict pesticide use and to encourage the development of more environmentally sustainable chemicals, often in advance of EU-level policies. A good example is the Netherlands Multi Year Crop Protection Plan (1991–2000) which has significantly reduced pesticide use.

Many Dutch farmers have managed their transition to green agriculture smoothly. The Netherlands has introduced sectoral policies to improve the efficiency of energy consumption in agriculture. The incentives to increase environmental production methods have been driven by both public policies and market initiatives including growing consumer preferences to environmentally friendly products. A good example is the Dutch Horticulture Environmental Program which stimulated environmental awareness in the cultivation of flowers, plants, bulbs and nursery stock products. The program essentially requires producers to keep records on their use of crop protection products, fertilisers and energy. In addition, retailers have increasingly demanded the use of environmentally-friendly farming methods. These various factors have promoted the successful green agriculture transition in the Netherlands.

Green agriculture innovations for sustainable growths

Green agriculture has presented big innovation challenges. These include producing more food without relying on the use of chemical fertilisers and pesticides. New innovations include replacing environmentally-damaging products and industrial production systems with new agricultural approaches which will help to protect biodiversity and mitigate climate change as well as addressing customer preferences and livelihoods.

Chemical and pesticide developments have contributed to past intensive agricultural practices with serious environmental damages. Today the green agricultural challenges are more complex including climate change, global warming, sustainability, biodiversity and food security all needing to be addressed simultaneously. These have meant that new sustainable agricultural innovations must involve a complex decision-making process that balances the immediate concerns of feeding the world against future concerns of sustainability.

A good example of this delicate balancing act is the OECD 2009 Declaration on Green Growth. The Declaration suggested that sustainable economic and environmental growths should be the key future challenge for all countries. The Declaration was a key precursor to a larger OECD Green Growth Strategy being developed. Green growth is defined as a way to pursue economic growth and development, whilst preventing environmental degradation, biodiversity loss and unsustainable natural resource uses. It will aim to optimise cleaner green growths with more sustainable growth models (OECD, 2009).

The development of sustainable green agriculture innovations will be heavily influenced by government policies and strategies. These are normally tailored to each country and their specific national contexts. Public and private partnerships (PPP) will be important. A good example is the emergence of various PPP green agricultural projects in different countries. There will be opportunities for countries to learn from each other and share best practices. A good example is the OECD and non-OECD countries sharing their green agricultural best practices. These exchanges should help to spur international exchanges plus greater global research collaboration which should promote co-operations between the developed and developing countries. It will also help to promote the internationalisation of the green agricultural value chains across the world.

Experts have suggested that innovations could contribute to meeting the green agricultural challenges through four key pathways. These include new technology, farming system innovations, integrated green regimes and cross-cutting innovations.

New science and generic technologies which have significant transformation potential for green agriculture will include biotechnology and bioproduction plus ICT technologies. These technological innovations could improve the environmental performance of farming systems through innovations in engineering, information technology and biotechnology. Newer technologies can reduce known toxins in agricultural production and substitute these with safer alternatives. It will also help to protect ground or surface waters, conserve natural habitats, reduce nutrient loads in soils, lower gaseous nitrogen loss and reduce the amount of non-renewable energy used in the cropping cycle.

New farming systems innovations with green potential will lead to different ways of organising agricultural production. This may involve the use of specific technological innovations to improve on how to organise agricultural productions and marketing. Good examples include Organic farming, Integrated Pest Management and Rice Intensification Systems.

Integrated National Green Regimes will involve specific technologies or agricultural production systems operating as part of specific national or regional green agenda with strong government policy support. Good examples include bio-fuels in Brazil and organic states in India.

Cross-cutting innovations would include discussions by governments and key stakeholders on which market or policy-driven mechanisms are suitable to drive green agricultural innovations. These would include local agriculture practices which will create different conditions for cross-cutting innovations in each specific country.

These four key green sustainable agriculture innovation pathways will be discussed in more detail in the following sections, with international examples.

Green agriculture and biotechnology challenges

Biotechnology innovations may support green agriculture developments but its uses have been quite controversial. Biotechnology innovations will involve the use of living organisms and bioprocesses in agriculture and production. Biotechnology could support sustainable development by improving the environmental efficiency of primary agricultural production and farm produce processing. Biotechnological innovations might help to repair degraded soil and water in different countries. Good examples include the use of bio-remediation, which is the use of live micro-organisms to reduce contaminants or toxins present in soil to benign products. Biotechnology can help to improve crop varieties that would require less tillage, fewer pesticides and fewer fertiliser applications. These will help to reduce soil erosion, degradation and pollution. Genetic fingerprinting could be used on farm livestock so as to improve their management and disease resistance. Genetic engineering could be applied to rare species, such as wild fish stocks, so as to help their populations to grow and prevent their extinction. Industrial biotechnology could be applied to minimise GHG emissions from chemicals production.

Biotechnology is currently being used to develop new varieties of food crops that have commercially valuable genetic traits, including herbicide tolerance, pest and disease resistance, improved flavours and tastes of farm products, etc. There are a lot of debates globally on biotechnology applications, especially on genetically modifications GM crops. A few GM crop species have received regulatory approval in different countries globally. The large majority of GM plantings have been for cotton, maize, rapeseed and soybean. The debates on GM crops have involved food security, commercial interests, plus ethical and legal concerns. Many vocal advocates of GM and GE food have argued that biotechnology-related food research would be essential in feeding the growing global populations while keeping environmental concerns at the forefront. A good example is as the global population continues to grow to 9 billion by 2030, many experts have advocated that advanced biotechnology, including GM, would be essential for green agriculture globally. It can be used to raise crop yield ceilings and grow crops without excessive reliance on pesticides. It can be used to help farmers with less fertile lands, with GM crops that are resistant to drought and salinity. Many governments from developing countries have put forth similar arguments as their countries are facing strong pressures from rising population and food demands.

There are also many opposition groups arguing on the basis of ethics. Some NGOs and activists have said that the claims of biotechnology's potential to feed the world have come from strong commercial interests that have big commercial stakes in its future applications. Public attitudes to GM crops have also varied greatly across different countries. Public perceptions of the risks and benefits of GM food can also change very quickly based on their publicly perceived risks and benefits.

There are also strong national and regional differences to the acceptability of GM crops. In general, European markets have been fairly anti-GM crops and food. On the other hand the USA has been an active promoter of the GM technology both at home and abroad. GM cotton has been approved in India and adopted, though not without opposition. A number of African countries have also adopted GM crops, although some are more permissive than others.

Government policies and regulations have significant roles in making bio-technology something that is perceived as a win-win solution for sustainable agriculture. Supporting policies and regulatory frameworks is necessary to ensure that biotech applications meet stringent bio-safety and environmental standards. It is also important for policy-makers to ensure that the new biotechnologies are not monopolised by strong commercial interests. There is also a lot of work to increase public understanding of the potential benefits and risks of biotechnology A good example is whilst the EU countries have been importing large quantities of GM maize and GM soy products for animal feed from countries such as Argentina, Brazil and the US, the EU has continued to restrict the use of GM plant varieties in domestic agriculture.

Green agriculture digital and ICT transformations

Information and Communication Technology (ICT) has three distinctly different roles in green agriculture. The first is precision agriculture, which uses advanced digital technologies to collect and analyse data for variations in soil or climate conditions. These ICT applications help to guide the application of the right agricultural practices to the right places. The ICT agricultural technology used would include Global Positioning System (GPS) sensors, satellite images, big data systems and digital information management tools, etc. These should help to collect various key agricultural variables information for analysis, which will help to guide the application of flexible green agriculture practices to improve crop management. These should help to increase crop productivity, raise farm incomes, improve efficiency and raise sustainability.

The second role of ICT in green agriculture is the use of ICT platforms and processes to promote communication, information exchange and networking between farmers, farming organisations and agricultural businesses. Digital farming tools and applications are increasing being developed and applied on laptops, tablets or mobiles so farmers can use them efficiently. These applications have been helping the rapid distribution of farming information and agricultural knowledge to wider farming communities. A good example is the various new farming applications that have been developed in China on social media. These apps have been deployed on mobiles in China so that farmers in rural communities can now have access to up-to-date green agriculture information efficiently which they can then apply on their farms.

The third role of the use of ICT in green agriculture is the monitoring of land use patterns. Effective environmental databases can be used to track the status of various environmental indicators so as to improve sustainable environmental management. Key applications have included land cover assessment by counties, soil erosion, land use and cultivated land by slope and steepness and wetland inventories. GIS and satellite remote sensing have played important roles in the data collection. They have helped to pinpoint sensitive and vulnerable forests, watershed and fragile marine ecosystems which are of critical importance. They have provided essential information on both the quantity and quality of forest land, wildlife and marine resources. A good example is that China has created a range of national databases for land evaluation and management. These are helping them to monitor changing land uses.

GIS has also been instrumental in monitoring changes in forest land. A good example is that in Thailand, they have used satellite imaging in their Forest Loves the Water and Land project, to identify denuded forest areas in five northern provinces in Thailand.

ICT innovations can help green agricultural farmers to address environmental concerns with a high degree of precision and timeliness than was previously possible. Public investments in ICT applications will be important in coping with the unpredictability of climate changes. Appropriate government support with appropriate regulatory policy will help to ensure that sustainable land uses practices are followed.

Good country examples of ICT applications in green agriculture included both China and India, which have both been actively applying ICT. Both India and China have several ongoing programs applying ICTs to collect key agricultural data for evaluations and to support policy decisions.

China has been carrying out extensive water-mapping exercises in several villages with ICT applications. China has also created a range of national databases for land evaluation and management plus population, environment and sustainable development monitoring. New farming tools and applications have been developed and deployed on social media and mobiles in China so that farmers in rural communities can access the information efficiently.

In India, the Jal-Chitra software has been used to create an interactive water map of villages so as to enable communities to keep accurate water records. These include water availabilities from each source, recording water quality testing, listing water maintenance work, estimating water demands, plus generating monthly water budgets and monitoring rainwater harvesting systems. The new software has been tested in rural villages in Rajasthan in North India.

In the Philippines, the Manila Observatory has partnered with a mobile phone service provider, SMART, to provide telemetric rain gauges and mobile phones to farmers in disaster-prone areas. Local farmers can read their rain gauges and phone in their information to the observatory. The observatory can also send early warnings on storms and weather changes, via the mobile phones, to fishermen and farmers so they can take early action.

The International Crop Research Institute for the Semi-Arid Tropics (ICRISAT) integrated climate risk assessment and management system has been using remote sensing and GIS techniques to study rainfall patterns in Asia and Africa. They have then prepared suitable climate advisories for farmers in the drylands of Asia and sub-Saharan Africa to help them with green agriculture management.

Some countries have also been applying digital and IT big data system to monitor and reduce GHG emissions from their agriculture sector. A good example is the Holos model being applied in Canada. Holos is a Canadian government whole-farm modelling program which has been designed to estimate the GHG emissions based on information entered for individual farms in Canada. Holos will then estimate the carbon dioxide, nitrous oxide and methane emissions plus carbon storage and loss. The main purpose of Holos is to envision and test possible ways of reducing GHG emissions from farms in Canada. These have helped Canadian farmers to reduce the carbon footprints of their cattle by 15 per cent per pound of beef recently. Similar agriculture climate models should help other countries to better monitor and reduce their GHG emissions from farms (Agriculture and Agri-Food Canada, Holos Climate Change Agriculture Model, 2018).

Green agriculture and integrated pest control developments

Good pest controls and management are critical for successful agriculture globally. Integrated Pest Management (IPM) is a new ecologically-based green agriculture approach. IPM focuses on long-term solutions to pest control through a combination of techniques such as biological control, habitat manipulation, modification of agronomic practices plus use of resistant varieties. IPM will normally involve the application of a whole suite of different techniques so as to reduce the use of chemical pesticides. These could include crop rotation, monitoring the presence and growth stage of pests, use of antagonistic and parasitic organisms plus biological pesticides. IPM is usually more complex for farmers to implement as it will require different skills in pest monitoring and understanding of pest dynamics, as well as having to gain the cooperation of all farmers in a given area for effective IPM implementation.

IPM has been deployed in two types of farming systems. The first is in developing countries where labor is relatively abundant and where pesticides are expensive as well as environmentally-damaging. IPM has also been deployed in high value export horticulture. The stringent compliance requirements for pesticide residue norms in the European horticulture market have driven its successful implementation.

A good IPM example is the use of bio-control agents in the horticultural sector for the control of red spider mite by the Real IPM Company. The Real IPM Company is a Kenya-based company which has been commercialising biological control pest agents for the horticultural industry. It was started in 2000, when Kenya's largest horticultural and floriculture exporter, the Flamingo Holdings Group established Dudutech as a subsidiary to develop biological controls systems to reduce pesticide uses in the horticulture & floricultural sectors. Dudutech was established as a response to regulatory requirements in its major market of Europe where there was both a need to reduce pesticide residues plus also human rights issues associated with exposing workers to these during application. Two of Dudutech's key personnel established Real IPM in 2004. Their vision was to provide practical, sustainable and affordable reductions in pesticide use for both large-scale commercial growers and small-scale subsistence farmers throughout Africa and elsewhere. Real IPM's niche was providing a comprehensive suite of training and consultancy packages aimed at bringing clients up to speed on all aspects of best practice in sustainable pest and disease management programs. These programs are in compliance with the regulatory regimes governing the imports of fresh farming produce into the EU which include food safety and pesticide residues, etc. The Real IPM company also produces and sells seven biological control agents to deal with a range of crop pests.

A good case of IPM research and application is in India. The Indian government has provided financial assistance to state agricultural universities and other research organisations to research on develop biopesticides and biocontrol agents as part of its IPM strategy. A number of biopesticide production units and plant protection clinical centers have been established in recent years. As a result, the use of biopesticides and biocontrol agents in India has been rising. A key reason is that biopesticides are cheaper than chemical pesticides in India, as well as being eco-friendly and less risky in terms of resistance development.

An alternative to IPM is Non-Pesticidal Management NPM. While IPM can still use pesticides as a last resort, NPM advocates eliminating pesticides altogether. Non-pesticidal management relies on the farmers' knowledge and skills plus all working together as a community. It looks at the pest complex as a whole, rather than at individual insects. Farmers have to understand the many factors that influence pest numbers in their fields. These would include the life cycles of the insects, the incidence of pests and diseases, predator-prey relationships amongst different creatures, the relationships between growing monocrops and the pest population, and the management of soil fertility. Non-Pesticide Management will apply different techniques including deep ploughing in the summer to expose insect pupas, the use of light-traps and bonfires plus setting pheromone traps. NPM also uses biological pesticides, including chili and garlic extracts etc.

A good NPM example can be found in India. The Centre for Sustainable Agriculture in Hyderabad, India has had good successes in helping cotton farmers in Andhra Pradesh to switch to the more sustainable non-pesticide management system with the help of an NGO called SECURE which stands for Socio-economic and Cultural Upliftment in Rural Environment. Working together with NGO activists, Indian scientists were able to convince reluctant farmers to switch off pesticide uses through on-farm demonstrations. As the farming yields for farmers who have switched to NPM grew, the project began to take off with more farmers joining in. An approach that started as a demonstration project in one small village is now part of an official state policy package, with the Andhra Pradesh government rolling out the project in 11 districts.

Green agriculture systems intensification developments (SRI)

Important new green agricultural innovations to enhance sustainable crop productivities are being developed. A good example is the innovative Systems of Rice Intensification (SRI) that has been applied in over 25 countries globally. SRI is an innovative methodology for growing rice which was originally developed in Madagascar. SRI applies green agriculture approaches which are radically different from traditional ways of growing rice. SRI has helped to generate large environmental benefits such as reducing water consumption and lower fertiliser usages. SRI has involved the careful transplantation of single young seedlings instead of the conventional method of using multiple mature seedlings from the nursery. SRI then spaces the rice plants more widely and does not depend on continuous flooding of rice fields. SRI rice fields have used less seed and lower chemical inputs. These have promoted soil biotic activities in, on and around plant roots. These are then enhanced through liberal applications of compost plus weeding with a rotating hoe that aerates the soil. SRI applications in several countries around the world have shown good results with considerable increased rice yields. In addition, SRI has helped to produce rice crops that take a shorter time to mature and are drought-resistant. The innovative SRI approaches have also been extended from rice to other food crops such as wheat, finger millet, maize and kidney beans.

SRI has helped to increase rice yields on rural farmers' fields in over 25 countries. SRI, like IPM, has emerged from research and developments in green agricultural systems in developing countries. These developing countries have all been suffering from resource scarcity, lack of water and droughts. To date SRI has also been flourishing more in the green agriculture systems in developing countries. Despite its win-win characteristics, SRI has not spread widely into the more developed OECD countries, where its sustainability could be attractive. Some international agricultural research organisations and aid agencies are also sceptical of SRI despite increasing evidence that the SRI methods have helped to raise agricultural productivity.

A successful SRI green agriculture and social developments case study can be found in India. SRI's uptake and spread in India has been an unconventional one. SRI adaptation resulted from a civil society-led initiative, rather than from a more traditional research initiative or from government policy directive. In India the SRI methodology evolved quite independently of governmental policies and private sector involvements. Indian civil societies, inspired by stories of SRI's success elsewhere in the developing world, encouraged poor farmers to experiment with the SRI methodology on a pilot basis in different Indian states between 1999 and 2003. There were initial indifferences from formal Indian research organisations. The initial scattered SRI farm experiments grew into a wider green agriculture movement across India as more civil societies and farmers became involved.

Government officials from various departments such as irrigation and rural development also became involved. Women and child welfare groups also helped to spread the word across India. To date, several leading Indian states, particularly the smaller remote states, have demonstrated good SRI successes in India. Small and marginal farmers have taken up SRI in many parts of India. The Indian farmers have seen SRI as a good opportunity to overcome local food security problems and to cope with the growing drought problems.

Green agriculture and organic food production developments

The share of green organic agriculture has been rising globally in recent years. These fast growths have been driven by wide consumer demands for healthier organic food in both developed and developing countries globally. There have been wide variations in organic green agricultural practices across OECD and non OECD countries. Many countries have encouraged their farmers to convert to organic green farming by providing policy support and compensations for financial losses incurred during their conversion.

Organic green agriculture was developed as a holistic, ecosystem-based green agriculture approach. It was conceived as an alternative to the ecologically unsound practices of conventional high intensity chemicals based agriculture. It is important to distinguish between the certified organic agriculture and green agriculture, which is practised in an organic way but without certification.

The growth of organic food sales was initially led by consumer demands from high income European countries but has spread globally. The European organic consumer market has become the largest in the world. Consumers have been favouring organic produce for a variety of reasons. These include benefits to health, environment, plus perceived improvements in food quality and taste. There is now good accessibility to fresh organic produce in many supermarkets globally. Recent food safety problems in some countries, including BSE and foot and mouth disease, have also helped to drive up demands for organic food. In addition, the rising concerns by consumers about genetically modified (GM) foods and crops have also boosted demands for organic produce.

In most OECD countries, many organic farming information and standards are in place. A good example is that there has been extensive organic certification and labelling guidelines put in place by government agencies plus organic food suppliers and retailers. These are mainly in response to the large consumer demands as well as to aid consumer choice and purchase of organic food in the supermarkets.

In most countries, large consumer and market forces have driven the development of the organic green agricultural sector. A number of governments, mostly in Europe, have also offered financial incentives to farmers to convert to organic agriculture on the basis that these would help to generate more environmental benefits in their countries. There have also been some shifts in publicly-financed agricultural research towards green organic agriculture systems. A good example is that in a few countries the government central procurement policies have been changed to include the purchase of organic food by public institutions, hospitals and government departments.

Some other countries have used a carrot and stick approach on their farmers to encourage transition to organic farming. A good example is in Austria where the farmers have to sign a strict five-year contract and agree to comply with certain measures of the Austrian organic program in return for government fiscal and cash benefits. Generally such government financial incentives have been higher than other existing government agricultural supports. These have helped to ease the costs of transition to organic production by farmers.

Organic green agriculture in developing countries has also been rising rapidly. This was driven initially by rising organic farming exports to Europe and the West. Recently there have been rising domestic demands, especially from the rising middle class, in most developing countries. Agricultural economic studies have shown that organic farmers could earn much higher incomes than their traditional farming counterparts. A very important feature of organic green agriculture is that it is well-suited for small-holding farming. It builds on traditional knowledge and farming systems already being practised, in both the developed and developing countries.

A good example is that in India, a committee set up by the federal Planning Commission in 2000, recommended organic farming as the most viable option for the states in Northeastern India. Following the report, the Indian states of Sikkim, Mizoram and Uttaranchal have declared themselves to be Organic States. They jointly declared that they would all be committed to switch to an official agricultural policy which would support and encourage organic smallholder farming. The Organic States have been actively supporting their farmers in adapting to organic farming practices. The Organic State government agencies have facilitated the certification of organic products. They also helped farmers to build connections to international export markets. The Organic States have also invited donor funds from government and international agencies to help capacity building amongst farmers etc.

Green Conservation Agriculture developments

Green Conservation Agriculture was first developed in Brazil in the 1970s. It was based on the application of three major principles of agro-system management including minimal soil disturbance, protection of the soil through the permanent maintenance of plant cover at the surface plus the diversification of rotations and intercropping. Since then, green conservation agriculture has been developing rapidly in various countries, such as France. Today conservation agriculture has involved no-tillage techniques with minimal soil disturbance plus the maintenance of plant cover with a diversification of rotations and intercropping.

The diversity of production conditions and of farmer needs has led to considerable diversification of practices in green conservation agriculture globally. Conservation agriculture could involve a family of cropping systems rather than a single crop system. In some cases, seeds are sown directly through the crop residues whilst in others, the soil is still lightly prepared to facilitate crops installation. In general, conservation agriculture will involve changes in soil tillage techniques together with the use of cover crops and intercropping etc.

A good country example of green conservation agriculture is France. Although no-tillage techniques had been practised in France since the 1990s, it was only recently that conservation green agriculture started to be widely practised in France. Inspired by the Brazilian experiences, a group of French farmers applied green conservation agriculture techniques to their farms. These practices started to spread through informal networks amongst farmers, which helped to share information on production and cropping practices. The learning process, initially focusing on equipment and soil, gradually shifted to the use of cover crops. There has been a general shift away from no tillage practices towards conservation agriculture through socio-technical networks. These networks helped to combine a number of objectives and stakeholders, through multiple clusters, associated with various technical, agronomic and environmental aspects.

Conservation agriculture has also been tested in organic farms in the Rhône-Alpes region of France. However, it is more difficult to reduce soil tillage in organic farming, with two technical obstacles involving weed control and nitrogen nutrition. The use of leguminous cover crops to fix nitrogen is of particular interest in organic farming. Given the difficulty of controlling weeds mechanically, farmers and researchers are going through a process of determining whether the environmental conditions in organic farming could lead to a spontaneous change in the flora or whether the presence of a mulch or cover crop could modify weed emergence.

Another good example in green conservation agriculture is Argentina, where they have successfully applied zero tillage techniques. These zero tillage agricultural techniques have allowed farmers to sow seed into the ground with minimum soil disturbances. Agriculture in Argentina underwent a boom in the 1950s with large commercial cultivation of crops such as soybeans. Heavy applications of chemical fertilisers and pesticides boosted production rapidly until 1990s. These double-cropping farming practices have taken a serious toll on soil fertility as well as causing serious erosion and loss of organic matter, etc. Crop production began to dip even in resource-rich areas of Argentina. As awareness of the effects of inadequate soil management practices grew, so did interest in new and improved crop management techniques in the late 1990s. The government of Argentina encouraged research on the issue and commissioned studies to learn from other countries. They also organised numerous study tours for researchers and farmers. The active building of an informal network of policy-makers, researchers, farmers around common issues have contributed to a successful switch to zero-tillage farming practices in Argentina. Stakeholders worked together to overcome various hurdles which had inhibited wide adoption of zero tillage technology, including the absence of suitable weed control alternatives plus farmers' lack of funds and access to loans, etc. Apart from significantly improved yields of soybeans since the adoption of zero tillage technology, Argentina has reported lower rates of depletion of organic matters and higher soil moisture-holding capacities. These have led to reductions in soil degradations and increased productivity. Conservation agriculture and zero-tillage technology have now been widely adopted across Argentina. These have generated both positive economic and environmental outcomes with active buy-in by farmers across Argentina.

Green agriculture natural resource management developments

Integrated natural resource management INRM has become an important part of green agriculture globally. INRM is a systematic process to optimise agricultural natural resources. These include the multiple aspects of natural resource uses, such as biophysical, socio-political, and economic, so that these will meet the different goals of producers, users and community. The goals include food security, profitability, risk aversion, poverty alleviation, welfare of future generations, environmental conservation, etc. INRM focuses on sustainability as well as including all possible stakeholders from the planning level so as to reduce future possible conflicts. The conceptual basis of INRM has evolved in recent years through various research covering sustainable land use, participatory planning, integrated watershed management, and adaptive management, etc.

One INRM program is the new 'reducing emissions from deforestation and forest degradation' (REDD) mechanism. This is a new green agriculture and climate mechanism to slowdown the loss of forests globally. It involves paying forest countries to stop deforestation from 2013 onwards. REDD was expanded to REDD-plus in 2008 to account for measures to conserve sustainably managed forests, forest restoration and reforestation. REDD involved the transfer of funds from rich countries to poorer countries to slowdown deforestation. A mixture of government and private sector funding are expected to cover various costs. To measure a REDD project, it will be necessary to calculate the amount of carbon stored in the forest in question and then predict how much carbon could be saved by slowing down deforestation.

Another INRM program is Silvopasture, which has combined forestry and grazing of domesticated animals in a mutually beneficial way. The advantages of a properly managed Silvopasture operation include enhanced soil protection and increased long-term income, with the simultaneous production of trees and grazing animals. The Regional Integrated Silvo-pastoral Approaches to Ecosystem Management Project was an initiative funded by GEF. Its partners include CATIE, FAO and others. The project's objective was to assess Silvopastoral or forest grazing systems, to reha-bilitate degraded pastures so as to protect soils, store carbon, and foster biodiversity. It also developed incentives and mechanisms for payment for ecosystem services. These would result in benefits for farmers and distil lessons for policy-making on land use, environmental services and socio-economic development.

A successful example of Silvopasture application is in Latin America. From 2003 to 2006, cattle farmers from Colombia, Costa Rica and Nica-ragua, participating in the project received between US$2000 and US$2400 per farm. This represented 10 to 15 per cent of net income to implement the program. It led to a 60 per cent reduction in pasture degradation in the three countries. The areas of Silvopastoral land use, with improved pastures with high density trees, fodder banks and live fences, increased significantly. The environmental benefits included a 71 per cent increase in carbon seques-tration plus biodiversity benefits with increases in bird, bat and butterfly species together with a moderate increase in forested area. Milk production and farm incomes also increased by more than 10 and 15 per cent respec-tively. Herbicide uses dropped by 60 per cent which reduced chemical pol-lution. The traditional practice of using fire to manage pasture has become less frequent, which reduced GHG emissions significantly. Other demon-strated environmental benefits of the Silvopastoral systems include the im-provement of water infiltration; soil retention; soil productivity; and land rehabilitation. It also generated reductions on fossil fuel dependence by sub-stitution of inorganic fertiliser with nitrogen fixing plants. The project has successfully demonstrated the effectiveness of introducing payment incen-tives to farmers and integrated ecosystem management on the restoration of degraded pastures.

Green agriculture water management systems improvements

The agriculture sector is a major user of water in many countries globally. Agriculture globally has accounted for about 70 per cent of the world's freshwater withdrawals and over 40 per cent of OECD countries' total water withdrawals. Looking ahead, as the global population continues to grow, agriculture will have to produce almost 50 per cent more food by 2030 and then doubling the global food production by 2050. These increased agricul-tural productions will need to be achieved with better water management. Hence it will be important in future for farmers to raise water use efficiency and improve agricultural water management, while preserving the aquatic ecosystems.

The sustainable management of water resources in agriculture will in-volve water managers and users working together to ensure that water resources are allocated efficiently and equitably plus used efficiently. These will include irrigation with smooth water supply across the production seasons; water management in rain-fed agriculture; plus management of floods, droughts and conservation of ecosystems.

A good country water management case study is found in Israel. Israel has applied integrated water management systems to successfully manage their scarce water resources. Israel's agricultural sector has been applying intensive production systems due to the scarcity of natural resources, especially water with a climate that is largely arid and semi-arid. A narrow coastal strip and several inland valleys represent most of the fertile areas in Israel. Water supplied from aquifers and the Sea of Galilee makes irrigation possible. Saline seawater is used extensively and advanced technologies have been employed to optimise use of available water. Most of the water resources are in the northern and central parts of Israel but agriculture is being undertaken in the arid south and east. This reality has necessitated construction of an integrated water supply system to deliver water from north to south.

Israel has invested heavily in water and irrigation research since the 1950s. Early research indicated that water use was more efficient in pressurised irrigation than in surface irrigation. An irrigation equipment industry was established to develop drip irrigation automatic valves and controllers, media and automatic filtration, low-discharge sprayers and mini-sprinklers, compensated drippers and sprinklers. Fertigation is applied in most of the irrigated areas together with soluble and liquid fertilisers compatible with fertigation. Most of the irrigation is controlled by automatic valves and computerised controllers. With the land being divided into plots with harsh topographical conditions, only limited areas can be irrigated by mechanised systems, such as pivot irrigation.

These water constraints and varied climate conditions have stimulated the development of unique agro-technologies which allow the use of marginal water, such as brackish and reclaimed water. Brackish water has been used for irrigation of salinity-tolerant crops like cotton. In several crops, such as tomatoes and melons, brackish water has improved produce qualities, though with lower yields. The use of reclaimed water for the irrigation of edible crops has required a high level of purification. For that purpose, farmers in Israel have used a unique technology, the Soil Aquifer Treatment (SAT), in the densely populated Dan region. After tertiary purification, the water percolates through sand layers, which served as a biological filter, into the aquifer. From there water is pumped out at nearly potable quality and can be used for unrestricted irrigation.

Israel has also installed a countrywide network of agro-meteorological stations to deliver real-time digital weather data to farmers. These data have helped to adjust the irrigation regime. Diverse soil-moisture monitoring devices, including tensiometers, pressure chamber systems, and electrical resistance sensors, have been used for more precise specific adjustments.

Vegetal indicators, such as leaf water potential and fruit growth rates, have been used to improve precisions in water applications.

One of the main methods currently used in intensive farming is the closed water system. The unique feature of this system is the constant flow of water from the reservoir, through the covered breeding ponds, and back to the reservoir. In this case the reservoir would also serve as a bio-filter, reducing the concentration of nitrogen in the water, which is directly absorbed by the algae and bacteriologically broken down. Due to the high density of fish in the breeding ponds, farmers would enrich the water with oxygen and feed the fish protein-rich food. The result is a 40-fold increase in production. Other closed water systems based on biofiltration units are also being developed. The result is that more fish can be produced with less water.

Another method is the utilisation of water in reservoirs intended for irrigation. The use of reservoir water for agriculture also contributed to water savings. Fish farming in the desert provided a long-term solution to the problem of increasing fish production in a small country with limited water resources. This is feasible due to the desert aquifers or underground water sources. Due to the lack of fresh water, fish farmers have begun to exploit the sea. One method involved offshore cages along the coasts of the Mediterranean and the Red Sea. Another method involved breeding ponds located near the sea, which utilise seawater in a closed water system. The water is circulated from the ponds to the sea and back again.

Looking ahead, the expanding urban population in Israel, as well as potential political developments, will likely further reduce the available fresh water supply for agriculture. Potential solutions for Israel could be in the desalination of brackish water and higher water reclamation. In addition more parts of the annual crops could be grown under cover, where recycling would become routine. These green agricultural and water management innovations have helped Israel to develop a thriving agriculture sector despite their natural resource limitations.

Green urban agriculture developments

Green urban agriculture and peri-urban agriculture have been growing fast in recent years. Urban agriculture has been providing up to 15 per cent of the world's food requirements. In addition, up to 70 per cent of the urban households in developing countries have been participating in urban agricultural activities. These include producing vegetables, fruits, mushrooms, herbs, meat, eggs, milk and even fish in urban gardens and farming plots situated in private backyards, schools, hospitals, roof tops, window boxes and vacant public lands. In addition, urban agriculture has also generated various micro-enterprises in cities, such as compost production, food processing and sale, etc. These have helped to create new urban employments and contribute to sustainable urban economic developments.

Green urban agriculture has also contributed to the greening of cities by improving air quality and lowering city temperatures plus improving living conditions. There are also some key constraints to be addressed, including the lack of access to water, competition for land, food safety and hygiene, etc.

Many urban and peri-urban farmers, especially those in OECD developed countries, have used bio-intensive methods. These included companion planting, double-digging, solid and waste recycling, open pollinated seeds, composting, etc. Vertical farms and stacked greenhouses have incorporated water systems that would produce potable water from waste water and recycle organic waste back to energy and nutrients. In several OECD countries, urban and peri-urban production has also been linked to the expanding farmers' market movement. Urban agriculture has also helped to reduce the carbon footprint of agriculture as transportation to markets would be shorter and generate less GHG emissions.

A good example of green urban agriculture is found in Senegal. The government of Senegal launched in 1999 the micro-gardens project in Dakar. It has a dual purpose of generating incomes for the poor farmers in the city who had no access to farming land and providing fresh vegetables to poor families which would improve their food supply. These urban agriculture developments also contributed to a cleaner, greener city. The project has trained 4,000 families in micro-garden techniques. These involved training and helping them to access equipment plus marketing their produces. The project benefited from R&D by the Horticultural Development Centre (CDH) of the Senegalese Institute of Agricultural Research (ISRA). Despite city planners not having created enough spaces for urban farms, some city halls, schools and hospitals have made their backyards available to micro-urban gardeners. The micro-gardens project has also established sales outlets in all the regional cities. Annual yields have increased and costs have reduced through the use of alternative materials and drip irrigation kits. The project has also been collaborating with Italian NGOs in Dakar to establish a dedicated supply chain mechanism for micro-gardeners' produce. Television programs and advertisements have also helped to promote the micro-garden produce. The project has been working on developing special branding for their urban farm produce to promote further sales.

Another good example of urban agriculture is in India. India has a wide array of urban and peri-urban agricultural initiatives. In Mumbai, rooftop gardens have used recycled waste, compost and garden soil to produce fruits and vegetables and even cereals. Urban aquaculture has also been practised in Kolkata's Salt Lake marshlands, where local fishing cooperatives have been leasing patches of the wetlands.

In the USA, it is increasingly popular to use rooftop gardens to grow food in cities. Rooftop gardens have been a growing phenomenon in New York City and other major US cities. Some US NGOs have been offering free seedlings and supports. Another NGO, Just Food, has been offering courses on growing and selling urban grown food.

In Europe, there have been rising demands for good locally-produced food. There are also concerns that the transportation of agricultural produce over long distances has created large carbon footprints. Food miles is an increasingly used term that refers to the distance that food has to be transported from the farms until it reaches the supermarkets and then the consumers. Many companies have used food miles as a factor to assess the impact of food on environmental and global warming. A good business example is that Wal-Mart has adopted the use of food miles and has been using it as part of their logistic planning and profit-maximising strategy whilst reiterating its environmental benefits as well.

There are new interesting researches into urban waste recovery and green agriculture. An interesting example is the research on recovery of oil from waste coffee beans from urban cafes. Rising urbanisation and the urban coffee culture have led to many coffee cafes being established in urban cities by different coffee chains. These coffee shops all generate a large amount of coffee bean waste daily from their coffee sales. New eco-startups have been established with innovative green urban coffee waste recycling models. The new eco-startups would collect waste coffee bean from the various coffee stores with a win-win waste recycling agreement. Then they would recover useful oil from the waste coffee bean with various innovative extraction processes. The recovered oil would then be used in the cosmetic sector to replace palm oil or in the transport sector as a biofuel. These should reduce urban wastes whilst optimising palm oil and biofuel productions internationally. The combined efforts should help to reduce global GHG gas emissions and promote sustainability globally (Efthymiopoulos, 2018).

Green agriculture clean renewable energy developments

Green agriculture has great potential to contribute to generating clean renewable energy, especially biofuels. On the other hand, agricultural production also uses a significant amount of energy. So renewable energy for agriculture is an important new green energy option. It can help to reduce energy consumption and lower GHG emissions from agriculture.

The Integrated Food Energy Systems (IFES) have been developed to address these issues by simultaneously producing food and renewable energy. There are generally two main approaches. The first approach would combine food and energy crops on the same plot of land, such as in agro-forestry systems. A good example is the growing of trees for fuel wood and charcoal. The second type of IFES is the use of by-products or residues to produce useful produces. Good examples include producing biogas from livestock residues, animal feed from by-products of corn ethanol, or bagasse for energy as a by-product of sugarcane production.

A good example of IFES is biogas in Vietnam. Vietnam has embarked on an integrated land management scheme, after the government granted land rights to farmers. This was supported by the Vietnamese Gardeners'

Association (VACVINA). They tried to combine gardening, fish rearing and animal husbandry to optimise land use. Traditional fuels such as wood and coal for cooking have become increasingly scarce and contributed to deforestation. The project adopted an innovative IFES approach to address energy and environmental concerns simultaneously. Biogas digesters would use animal waste to generate bioenergy, whilst the slurry would be used as a fertiliser to improve soil quality. A farmer must have at least four to six pigs or two to three cows which would provide the required amount of animal dung feedstock. They would pay the total installation cost for the digesters to local service providers. The farmers would then operate the biodigester, which would normally produce enough daily fuel for cooking and lighting.

Integrated National Green Regimes (INGR) has also been used to promote renewables developments. In INGR, specific agricultural production systems would be operated in line with the national green agenda. A good example is biofuels production in Brazil. Brazil has become the world's largest biofuel market, in line with their government's national green agenda. The Brazilian Government launched its National Alcohol Program, Pró-Álcool, in 1975 as a nation-wide program financed by the government to phase out automobile fuels derived from fossil fuels in favor of ethanol produced from sugar cane. Brazil's 30-year-old ethanol fuel program is based on the most efficient agricultural technology for sugarcane cultivation in the world. It uses modern equipment and cheap sugar cane as feedstocks. Brazilian ethanol from sugarcane is a renewable fuel that is cost-competitive with petroleum fuels for transport applications in Brazil. Ethanol production is more economical in Brazil than in the United States due to several factors. These include the superiority of sugar cane to corn as an ethanol feedstock. Brazil's large unskilled labour force has supported the labour-intensive sugar cane production requirements. Whilst the USA and Brazil have produced about the same volume of ethanol, the USA have used almost twice as much land to cultivate their corn for ethanol as Brazil did to cultivate sugar cane. The residual cane-waste, bagasse, is used to generate heat and power. This has resulted in very competitive overall pricing and efficient energy balances. There are currently no longer any light vehicles in Brazil running on pure gasoline. Since 1976 the Brazilian government has made it mandatory to blend anhydrous ethanol with gasoline, up to 22 per cent. In Brazil, sugar and ethanol are being produced on an integrated basis. The option to produce more or less of each product is influenced by the relative pricing. Some experts have suggested that the successful Brazilian biofuel ethanol model is sustainable only in Brazil due to its specific conditions and the enormous amount of arable land available. Brazil's ethanol infrastructure model has also required huge taxpayer subsidies and government supports over decades before it could become viable.

Agriculture bio-feedstock and bio-chemicals innovations

The use of renewable bio-feedstocks from agriculture and waste sources to make biofuels and bio-naphtha is being developed with various innovative processes. The use of biomass for bio-naphtha or biochemicals production will help to reduce fossil fuel consumption whilst supporting green chemicals production. It will have a lower carbon footprint compared to fossil-based petrochemical production.

Bio-naphtha is essentially a paraffinic hydrocarbon which is similar to the traditional fossil-based light naphtha. Bio-naphtha can be processed in a conventional refinery or petrochemical complex into a whole range of bio-chemical products including bio-ethylene, bio-propylene, isomerate, reformate, light olefins or aromatics. These can then be processed into bio-plastic and biopolymer products.

New innovations are being developed for the production of bio-naphtha from various wastes including agricultural biomass, food wastes and renewable wastes. Some of these processes involve the combination of the bio-refining together with traditional steam cracking. Bio-refining catalytically converts the triglycerides and fatty acids from fats, algae and vegetable oils to high quality synthetic paraffinic kerosene (SPK), biodiesel and bio-naphtha in three steps. First, the raw feedstocks are treated to remove catalyst poisons and water. In the second step, the fatty acid chains are deoxygenated and transformed into mainly paraffins in a hydrotreater. For most bio-oils, fats, and greases, the hydrotreater liquid product would consist mainly of C15 to C18 paraffins. In the third step of the process, these long straight-chain paraffins are hydrocracked into shorter branched paraffins. The hydrocracked products would fall mainly in the kerosene and naphtha boiling ranges There are also high temperature thermo-chemical processes to convert renewable bio-feedstocks to biofuels or biochemicals such as via pyrolysis and gasification or low temperature enzymatic processes. However these are currently more expensive and not cost-competitive with conventional fossil-fuel based processes. Another new process would involve the hydrolysis of biomass to oligosaccharides using a benign acid. In the second step, the oligosaccharides are deoxygenated with hydrogen to form light naphtha with C5 to C6 paraffins. The product is separated using a phase separator and the water recycled to the hydrolysis reactor.

A good bio-naphtha business example is the new Ikea Neste commercial pilot plans. In follow-up to their new partnership established in 2016, Finnish oil producer Neste and the Swedish retailer Ikea have announced plans for commercial-scale pilot production of bio-based polypropylene (PP). This will be the first large-scale production of renewable PP globally. They would also be able to produce bio-polyethylene PE. The PP test plant is planned to be started up in 2019–2020 and will use bio-naphtha feedstocks. This will be a byproduct generated in the production of renewable diesel from agricultural waste and used cooking oil at the Neste biorefinery in

Rotterdam, the Netherlands. The bio-naphtha will be processed in a chemical cracker to make bio-ethylene C2 and bio-propylene C3 streams which could have a renewable content of as much as 50 per cent. The bio-ethylene and bio-propylene grades produced in the pilot facility can be processed with the same machinery converters use to manufacture conventional plastics. Currently the cost of the bio-naphtha is double that of the fossil fuel-based naphtha. Further developments will be necessary to improve the process efficiency and economics.

Initially, Ikea plans to use the new bio-plastic in a few products in its current furniture ranges, such as storage boxes. By 2030, Ikea is planning that all their plastic products and furniture sold in its stores globally will be made from recycled or renewable plastic materials (Plasteurope.com, 2018).

Green agriculture policy support developments

Strong government policy support for green agriculture is a critical requirement for successful green agriculture. Governments have to develop win–win policies, with inputs from the agricultural sector, so as to improve environmental sustainability and agricultural productivity plus sustainable economic growths. Policy-makers will need to consider the various potential benefits against trade-offs on a holistic basis. In addition, win-win options may be highly contextual and specific for different countries. A few good examples from different countries will be discussed below.

One of the best examples of green agricultural policy success is Brazil's biofuel. Brazil's 30-year-old ethanol fuel program has been developed using some of the most efficient agricultural technology for sugar cane cultivation in the world. It has been using modern equipment and cheap sugar cane as feedstocks. The residual cane-waste, bagasse, has also been used to generate heat and power. These have resulted in very competitive ethanol biofuel pricing with efficient energy balances. Brazil's biofuel ethanol model and infrastructure have also required strong government support and huge taxpayer subsidies over decades before it could become viable. Experts have suggested it would be difficult for other countries to copy Brazil's biofuels policies due to Brazil's special conditions.

Policies supporting conservation agriculture and zero tillage technology applications have been successful for countries such as Brazil and Argentina. These have helped to reduce the threat of erosion by heavy rainfall and improved soil quality so that the soils sequester carbon. They have also significantly raised the incomes of farmers by giving them more adaptation options. In Argentina, farmers were included in the initial consultations and decision-making, which gave them a greater involvement and stake in the policy outcomes.

The integrated green policy approach adopted in the Netherlands has been a win-win approach which has well worked in the socio-political context. Policy-makers have been able to promote ecologically sound agricultural developments, without compromising on market competitiveness. These were made possible by a combination of appropriate policies and incentives. These included increased funding for research to develop a more preventive approach to crop protection and sustainable production plus incentives to industry to develop more environmentally-friendly pesticides. There were also policies to improve energy efficiencies in agriculture. The Dutch efforts at educating industries and consumers about environmentally-sound green agriculture have also helped industry to respond to consumer demands for green products.

Israel's water management system has been a win-win strategy for Israel. It has helped to provide Israel with food security and agricultural production self-sufficiency as well as a strong export industry. The Israeli government water policies have maximised the use of its scarce water resources to support its growing agricultural sector. It also had profound ecological effects on greening the country, conserving its scarce resources and rehabilitating soils.

The SRI has been a win-win green agriculture option adopted in developing countries with government policy supports. It has helped to increase rice yields, reduce water and fertiliser uses and has used fewer seed and chemical inputs. It is a green farming system innovation that has emerged out of civil society-led initiatives rather than from formal research. The key driver of this win-win green agriculture approach has been the need to cope with resource scarcity in some of the poorest developing countries with challenging agro-ecological conditions.

7   Carbon emissions trading systems management

种瓜得瓜
Zhòng guā dé guā
If you plant a melon seed, then you should get a melon when it is grown.
You reap what you sow.

Executive overview

The rising CO_2 and GHG emissions from different countries have contributed to worsening global warming and climate change impacts globally. Various improvements are being implemented to reduce CO_2 emissions so as to lower global warming. The deployment of different carbon emissions trading systems has been growing globally. These should help to control CO_2 emissions and reduce climate change impacts. Details of various carbon emission trading systems will be discussed in this chapter together with international examples.

Climate change and carbon emission trading system developments

Climate change and global warming have been described as two of the biggest problems confronting the world. CO_2 has been shown to be one of the key GHGs causing global warming and climate change globally. It is important to recognise that CO_2 has played an important role in the earth's climate. It is an important GHG which has helped to keep the earth's surface at the right temperature for life for many years. After the start of the Industrial Revolution, the earth's CO_2 concentrations started to rise. This was caused by the various industrial and human activities which generated massive amounts of new CO_2 and GHG emissions, particularly by the combustion of fossil fuels, especially coal and oil, associated with the various industrial, transport and residential activities.

Scientists have shown the concentrations of CO_2 in the earth's atmosphere have been rising exponentially, at rates of about 0.17 per cent per year, since the Industrial Revolution. These have been mainly caused by the combustion of fossil fuels, especially coal and oil. In addition, large-scale deforestation activities in various countries have also helped to reduce the earth's capacity to reabsorb CO_2 via photosynthesis. In 2015, global CO_2 concentrations rose passed 400ppm. This was more than 40 per cent higher than the pre-industrialisation CO_2 concentration of 280ppm. The current concentrations of CO_2 in the earth's atmosphere have been found to be higher than at any time over the last 800,000 years. A good evidence for this is that the air sampling station at the Mauna Loa observatory in Hawaii has recorded atmospheric CO_2 concentrations rising past 400ppm.

IEA The International Energy Agency (IEA) has in their latest Global Energy and CO2 Status Report of 2018 reported that global energy-related CO_2 emissions rose 1.7 per cent to a historic high of 33.1 Gt CO_2 in 2018. Emissions from all fossil fuels have increased globally, driven by higher energy consumptions which increased at nearly twice the average rate of growth. Higher electricity demand was responsible for over half of the growth globally. The global power generation sector accounted for nearly two-thirds of the GHG emissions growths. Coal used in power generation alone generated over 10 Gt CO_2 emissions, mostly in Asia. China, India and the USA. These contributed to some 85 per cent of the net increase in emissions, whilst emissions declined for Germany, Japan, Mexico, France and the UK (IEA, 2019).

In addition, CO_2 will persist for much longer in the earth's atmosphere than other GHGs. It can take hundreds of years for CO_2 concentrations to return to pre-industrial levels. A good example is that scientists have found some CO_2 in the atmosphere dating back to World War 1.

The rising CO_2 and GHG levels have contributed to the worsening global warming and climate change impacts globally. Experts have warned that climate change could cause global economic losses of US$1–4 trillion by 2035 and the resulting 'carbon bubble' could cause losses higher than the 2008 financial crisis. Many countries across the world will be suffering major climate-induced economic losses. The developed countries have been predicted to be amongst the biggest economic losers. Whilst there will be some significant losers, the overall impact on global GDP will likely be balanced. This is because the growth of low carbon economy in different countries will generate new business growths, new employments and sustainable economic growths. Good examples include countries like Japan, China and the EU, which have already seen major low carbon economic growths including green finance growths and rising employment in new green businesses (BBC Environment, Carbon 'Bubble' Could Cost Global Economy Trillions, 2018).

Climate change has already incurred major financial losses and social costs to different countries. These carbon social costs will vary from country to country depending on their specific social and economic conditions. The US Environmental Protection Agency (EPA) has estimated that the global average social cost of carbon would typically be about US$41/tonne.

The specific costs for different countries will be different depending on their specific conditions. Good examples include India's cost, which is about $86 per ton, the USA is close to $48, Saudi Arabia is $47 and China, Brazil and the United Arab Emirates are about $24.

The application of carbon emission trading schemes (CETS) has also been growing in different countries. These should help to lower carbon emissions and contribute to controlling overall global emissions. Appropriate carbon pricings have also been shown to have a marked impact on controlling carbon emissions. These will need government policy support for CETS with a level of price certainty provided. In order to encourage decarbonisation, a suitable carbon price floor should be considered. Globally carbon markets worldwide have raised close to US$30 billion by the end of 2016. These revenues can be reinvested in other climate change mitigation and adaptation programs. Most carbon markets have yet to yield significant emission reductions to date but these are likely to improve with more experience and learning. Key lessons learnt include better pricing and allocation plus improved management (Ricke et al., 2018).

To mitigate rising carbon emissions, the various new carbon emission trading systems (CETS) in different countries will have important future roles. Details of the key CETS will be discussed in this chapter, with relevant international examples.

Global carbon emission systems developments and challenges

CETS have been adopted by various countries. Carbon emissions trading is a form of emissions trading that specifically targets CO_2 management and trading across nations. CO_2 is calculated in tonnes of CO_2 equivalent or tCO_2e. These have constituted the bulk of carbon emissions trading globally. Carbon emission trading methods have been utilised by different countries in order to meet their Kyoto Protocol obligations. These would involve the reduction of carbon emissions so as to mitigate future climate change implications (IMF, 2008).

The current carbon emissions trading policies would allow legal entities, including countries, cities or companies, to buy or sell government-granted allotments of CO_2 output. The World Bank has reported that over 40 countries and 20 municipalities have been using either carbon taxes or carbon emissions trading. These have covered about 13 per cent of the annual greenhouse gas emissions globally.

Under the current CETS, various governments would distribute a finite number of government granted CO_2 credits to individual companies. These would become the carbon emission cap for these companies. The companies could only emit as much CO_2 as their credits would allow. Those companies which have been emitting below their CO_2 limit could then sell their CO_2 credits to other companies that have been exceeding their limits. This would then constitute the carbon trading part of CETS. The overall global goal would be to control and reduce the global carbon emissions so as to slow down global warming. Industries, like power generation utilities, would normally be the biggest CO_2 generators and traders (Amadeo, 2019).

Under the current global CETS, entities such as companies or cities, that have more carbon emissions, should be able to purchase the right to emit more from other countries or companies that have fewer emissions. In this way, the countries or companies emitting more carbon could then satisfy their carbon emission requirements without increasing the overall global carbon emissions. These CETS should helped to provide some of the most cost-effective carbon reduction methods and maximise carbon reductions for a given expenditure by a country or company (Smith, 2008).

In the 1980s, US President George H.W. Bush showed that the 'cap and trade' system would work well in the USA to control pollutant emissions. He used it to curb various environmental pollutants which have been causing acid rain. It was the first cap and trade program in the world. The global CETS market took off when the EU instituted a CO_2 cap and trade program in 2005. They set a cap on the total amount of CO_2 that heavy industries and utilities in EU could emit. In November 2017, the EU further reduced the carbon cap by 2.2 per cent each year through to 2030. The key goal of the EU cap would be to further reduce carbon emissions by 43 per cent across the EU by 2030.

The EU has been issuing about 2 billion European Union Allowances (EUA) each year. To comply with the EU mandate, EU companies could either take measures to emit only what they have been allowed to emit or re-duce their emissions below the allowed amount and sell or bank the surplus EUAs. If they continue emitting above their allowance then they have to buy EUAs in the marketplace to cover their excessive emissions.

Certified Emission Reductions Credits (CERC) could also be traded. These were created by the Kyoto Protocol. They are credits issued to pro-jects in developing countries that have led to emissions reductions. There are also greenhouse gas emission credits, which cover more pollutants than just CO_2. They can fulfil nation-specific caps in the USA, the UK, Canada, New Zealand and Japan.

The units which may be transferred under Article 17 emissions trad-ing, would each equal to one metric tonne of emissions (in CO_2-equivalent terms). These might be in the form of an assigned amount unit (AAU) is-sued by an Annex I Party on the basis of its assigned amount pursuant to Articles 3.7 and 3.8 of the Protocol. A removal unit (RMU) can be issued by an Annex I Party on the basis of land use, land-use change, and forestry (LULUCF) activities under Articles 3.3 and 3.4 of the Kyoto Protocol. An emission reduction unit (ERU) would be generated by a joint implemen-tation project under Article 6 of the Kyoto Protocol. A certified emission reduction (CER) would be generated from a clean development mechanism CDM project activity under Article 12 of the Kyoto Protocol. Transfers and acquisitions of these units would have to be tracked and recorded through the approved registry systems under the Kyoto Protocol.

The ability to buy and sell EUAs, CERs, and other units on a freely traded market has created a new form of carbon currency. Traders would include not only the emitters themselves but also banks, hedge funds, and other investors. They would provide liquidity and increase market efficiency. Emissions trading has also given polluters more incentives to reduce their emissions. However, allocating permits on the basis of past emissions ('grandfathering') could result in firms having strong incentives to maintain their emissions. For example, a firm that has successfully reduced its emissions might receive fewer permits in future rounds.

Carbon emissions trading has been steadily increasing in recent years. Carbon emission trading experts have estimated that the global market for carbon trading could be valued at over $170 billion. These carbon trading revenues could be used to be reinvested in suitable climate change mitigation and adaptation programs. Looking ahead, experts estimate that the global CETS market could grow further and might exceed $1 trillion by 2020. A major element is the EU's Emission Trading Scheme which caps emissions for any company doing business in the EU. However, various other countries have also introduced their CETS. Some of the key regional CETS schemes will be discussed below (Ricke et al., 2018).

Regional carbon emission trading schemes developments

Various countries have been developing their CETS as a means to control CO_2 and GHG emissions. Various CETS programs have been established and put in place in Europe, North America and parts of Asia, especially in China, India and South Korea. (Climate Policy Info Hub, 2018).

The world's largest carbon market is currently the European Emissions Trading Scheme (EU-ETS). It is covering the EU region, which has been emitting over 2 billion tonnes of CO_2 each year. The EU-ETS has been in operation since 2005. Currently, the EU ETS has established operations in 31 countries. These cover all the 28 EU Member States as well as Iceland, Liechtenstein and Norway. The EU ETS covers CO_2 emissions from emitters in the power sector, combustion plants, oil refineries plus iron and steel works, as well as installations producing cement, glass, lime, bricks, ceramics, pulp and paper. More than 10,000 industrial and utilities entities have been covered. These have accounted for around 2 gigatonnes or 40 per cent of EU total GHG emissions.

In North America, the USA and Canada have followed different development paths. The Federal Government of Canada has required all provinces to put a minimum price on pollution of $20 a tonne of emissions by January 1 2019. Saskatchewan, Manitoba, Ontario and New Brunswick have not complied and would have a federal carbon levy on fuels as well as a cap-and-trade system for large industrial emitters imposed on them. British Columbia, Alberta, Quebec, Newfoundland and the Northwest Territories have all put a price on pollution high enough to meet federal standards and the revenues in those provinces are being handled by those provincial governments. The Canada Federal Government forecasted that they will collect over C\$ 2.3 billion in carbon taxes from these provinces and 90 per cent of that will go to household rebates. The payments vary because carbon taxes collected will be higher depending on how provinces power and heat homes. The remaining 10 per cent will be handed out to small and medium-sized businesses, schools, hospitals and other organisations which cannot pass on their costs from the carbon tax directly to consumers (Alini, 2019).

In the USA, various national emission trading programs are still being discussed by the politicians on the federal level with different arguments. Experts advise that in reality the USA has already been suffering high climate-induced costs, as an invisible carbon tax. Good evidence of this is the USA National Climate Assessment (NCA) in 2018 has reported that climate change has had serious impacts on the US economy. They estimated the USA could suffer climate costs of \$500 billion per year due to hurricane damage, crop losses, coastal flooding and adverse health risks etc. Under the worse scenarios, climate-induced damage and costs could result in an annual 10 per cent reduction in the USA gross domestic product by 2100 (NCA, 2018).

At the regional level, various North American carbon trading systems have emerged between states. A good example is that nine US States have joined forces in a joint trading system called the Regional Greenhouse Gas Initiative (RGGI). The Canadian province Quebec has linked up with California's emissions trading program in January 2014. The ETS programs in California and Quebec currently cover sectors emitting nearly half a billion tonnes. The Regional Greenhouse Gas Initiative (RGGI) is a carbon market among states in the north-eastern and mid-Atlantic US that entered into force on January 2009. It covered only CO_2 emissions from electricity generation. The Western Climate Initiative (WCI) is an initiative of US states and Canadian provinces to jointly develop climate change policies (Coglianese & Nevitt, 2019).

In Asia there have been strong developments by various governments toward emission trading recently. A good example is South Korea's emission trading program, which entered into force in January 2015. Korea's ETS covered sectors that emit over half a billion tonnes of GHG. The Korean ETS began in January 2015 and covers over 60 per cent of the country's emissions. Over 500 companies are covered in the power and industry sectors, but also waste and domestic aviation (World Bank, State and Trends of Carbon Pricing, 2015).

In China, seven Chinese pilot emission trading systems have been running and collectively cover sources emitting over one billion tonnes of CO_2. These include five Chinese cities and two provinces which have started pilot carbon markets. These regions together would account for about one-fourth of Chinese GDP and CO_2 emissions. Chinese leaders have recently put forward a new national emission trading scheme which will be discussed in more detail in this chapter.

In other parts of Asia Pacific, the landscape is more mixed. Japan is not planning to implement a national emission trading system. However the Tokyo Metropolitan Government has been operating a trading scheme for indirect CO_2 emissions since 2010. The Tokyo Metropolitan Government Emissions Trading System (TMG ETS) has targeted energy-related CO_2 emissions from industrial facilities as well as public and commercial buildings in Tokyo. New Zealand's small CETS has been operating since 2008. The New Zealand Emissions Trading Scheme (NZ CETS) is the only CETS to include forestry as a covered sector. Australia abandoned a long-planned national ETS in 2013 after a change in government.

In Central Asia, the Kazakhstan Emissions Trading Scheme started with a pilot phase in 2013 covering CO_2 emissions. Energy, mining and metallurgy, chemicals, cement and the power sector are included in the scheme.

It is interesting to note that the various existing and new emission trading schemes are very different regarding design scopes or allocation methods. On sectoral coverage, the EU-ETS have been focusing more on industry and large energy producers. The new CETS in China is also focusing on their large power sector. Some of the emerging ETS schemes in Asia have also included smaller facilities, buildings and indirect emissions from energy consumption. On the type of targets, the ETS schemes in Europe have included caps whilst the Chinese pilot ETS have largely use an intensity metric for the required emission reductions. On their allocation method, only a few ETS worldwide have a significant share of auctioning from the beginning. Many ETS schemes have applied grandfather, such as the EU-ETS. Grandfathering is an allocation method by which emitters would receive permits on the basis of their past emissions from the beginning. The Chinese ETS pilots have allocated most allowances for free.

Looking ahead, the CETS in the emerging economies in Asia could eventually overtake the EU and other OECD countries as centers for emissions trading in future, based on their higher economic growths. These could significantly change the style, nature and challenges of the future international carbon market.

China carbon emissions trading system developments

China has just launched its new national ETS for the power sector after their ETS pilot trials. In 2011, the China National Development and Reform Commission (NDRC) announced seven official Emission Trading Scheme ETS pilot programs in five cities which included Beijing, Shanghai, Tianjin, Chongqing and Shenzhen plus two provinces which included Guangdong and Hubei. The initial implementation of the pilots started in 2013. In 2015 all the pilot programs had been established and started up. The experience from these pilots had supported the Chinese national government's plans to implement a national ETS. The Chinese pilot programs varied in terms of caps and targeted sectors. A good example is the Beijing pilot was the only pilot which required annual absolute emission reductions. The Beijing authorities have quantified the amount of CO_2 and other GHG emission reductions required for existing facilities in the manufacturing and service sectors in Beijing. The other ETS pilots have specified required reductions in emissions intensity which was expressed as a ratio of emissions per unit of product. The Shenzhen and Tianjin ETS pilots allowed individual investors and entities that were not covered in the ETS, such as financial institutions, to participate in emission trading. These had led to higher trading frequency and potentially larger price fluctuations (ICAP, 2014).

The emission allowances in the seven Chinese ETS pilots have generally been given out for free, based on grandfathering. For the power and heat sectors, benchmarking has been deployed. It was based on allocation methods by which emitters would receive allowances based on measurements relating to the sizes of their emissions. Two of the ETS pilots, Shenzen and Guangdong, had also started to experiment with a small amount of emission auctioning. All the Chinese ETS pilots were required to accept federally-approved offsets credits which were known as 'Chinese Certified Emission Reductions' or CCERs. These have mostly come from previous Chinese Clean Development Mechanism (CDM) projects under the Kyoto Protocol's CDM. Indirectly, this opened up a link between the pilots, as they have at least one type of compliance unit in common.

The first seven pilot markets covered 3,271 liable enterprises and 1,373 MT CO_2. These were equivalent to about half of the potential future China national market. Over time, trading activities in these pilots have grown. A good example was that the trading volumes and values grew 106 per cent and 29 per cent respectively in 2016. A high compliance rate has also been achieved with 99 per cent of included enterprises abiding by their pilot's rules. Some industrial leaders have emerged among the pilots which helped to establish best practices for the national market. Guangdong and Hubei have experimented with auctioning allowances. The Beijing market has maintained the highest and most stable carbon prices, which hovered around 50 yuan (US$7.6) per tonne. After the new national market begins to trade, these pilots would be expected to continue operating for some time so as to keep enterprises outside of the national market scope engaged on data collection and carbon reductions.

In 2018, China's National Development and Reform Commission launched the new China National Emissions Trading System (ETS) after approval from the State Council. The launch was in line with the commitment first made by President Xi Jinping in September 2015 during the bilateral meeting with USA. The launch of the national ETS gave a clear signal that greenhouse gas emitters in China would be held accountable for their carbon emissions and environmental pollution impacts. The ETS compliance cycle should help to control and reduce carbon emissions nationally across China. It should also create a valuable system for collecting company-level carbon emissions data which should provide a good basis for developing and improving future carbon policies in China.

The initial China national ETS would cover China's power sector. It would cover around a third of China's total carbon emissions which would be about 3.3 billion tonnes annually. These emissions would already be much larger, about 165 per cent, than the European Union's ETS system which covered 2 billion tonnes of emissions. The power sector is the most appropriate sector for China to start its national ETS as it has the most credible and transparent emissions data. That is also in line with best practices from both the EU ETS and the California carbon markets, which are the world's two largest and longest running emissions trading systems. Both of these have also included their power sector in their initial phases.

Looking ahead, it is planned that the China ETS will expand in future to cover the other key eight sectors of China's economy which will represent 40 per cent of the country's emissions. According to preliminary guidelines from the National Development and Reform Commission, the national emissions trading system would be extended in phases to cover each of the eight broad industrial sectors which will include power generation, petrochemicals, construction materials, chemicals, iron and steel, non-ferrous metals, pulp and paper plus aviation. China's new national ETS would also require careful monitoring and support so that it could be optimised and harmonised. In particular, there should be improvements to enhance the national ETS' legal foundation and allocation methodologies plus data quality.

Compared to other climate policies in China, experts have shown that carbon pricing would play the largest role in reducing China's GHG emissions in the future. By 2030, appropriate carbon pricings could reduce CO_2e by over 25 per cent based on climate modelling by the National Center for Climate Change Strategy and International Cooperation, the Energy Research Institute, and Energy Innovation. Their study results have shown ETS to have the biggest impact on emissions reductions amongst all different policies, including carbon taxes. The study also showed that other complementary policies must be implemented simultaneously for China to achieve its climate goals. China has plans for ETS to work alongside a slew of other climate policies and targets, which would include power sector reforms and new energy vehicle sales quotas, etc.

Careful market design would also be required to achieve the desired emissions reductions in China. Prices in China's pilot markets have fluctuated but have mostly stayed below 60 yuan/tonne (US$ 9/t) and averaged approximately 30 yuan/tonne (US$ 4.5/t). Results from the China Carbon Forum's 2017 China Carbon Pricing Survey showed that the market would expect the carbon prices to rise steadily in the future as the national market trading picks up. However there are big uncertainties on the extent of the future price risings which are currently difficult to forecast (China Carbon Forum, 2017).

The US Environmental Protection Agency (EPA) has estimated that the global averaged social cost of carbon would be US$41/tonne (272 yuan). Almost all ETS carbon prices worldwide would be below this price. The low prices seen across all different carbon markets have been largely due to over-allocation of allowances, which has also occurred in China's pilots. This global issue is starting to be addressed through price floors by different countries. Two good examples are that the UK and California have introduced these price floors. Similar policies would also be likely to be implemented in China in future.

The new China ETS is expected to have a real impact on corporate investment decisions in China. This was shown in the recent survey by China Carbon Forum which showed that carbon pricing would increasingly affect future corporate investment decisions in China. Economists had predicted that in the utilities and power generation industries, the cash flows of various coal-fired power generators would be adversely affected over the longer term as carbon emissions became more controlled and restricted. The earnings forecasts of some of the largest independent power producers in China could drop 5 to 35 per cent. However, under the emissions trading system, the renewable energy sector would be able to earn extra revenues through selling carbon credits to those industries which would be emitting more than their allowed quotas (China Carbon Forum, 2017).

The new ETS will have significant impacts on China's manufacturing industries. China's chemicals industry would likely face stricter controls as they have been generating significant environmental impacts. China's growing coal-to-chemical industry could be seriously affected as it has been using synthesis gas from coal as a major feedstock. CO_2 is generated and used at different stages of the MTO MTP polyolefins production process. Analysts have forecasted China would impose stringent environmental standards on coal-to-chemicals developments. In aviation, analysts predict that participation in the ETS scheme might be restricted to domestic carriers such as intra-China flights. However, the impacts should be minimal for airlines as they have already been driving fuel efficiency and emissions reductions. Over the longer term, this would prompt airlines to use more fuel-efficient aircraft and possibly use bio-jet fuels, as some international carriers have started to do.

For the construction materials industry, the ETS scheme should trigger a change in the Chinese construction methodology from being labour-intensive to more prefabrication and modular. These would be in line with international practices which should help to reduce wastage and increases efficiency. Prefabrication would include pre-assembling components in the factory and then transporting complete assemblies to the construction site. These should help to improve efficiency and reduce waste against the tradition method of delivering materials to site and then assembling onsite.

For the steel and non-ferrous metals sectors, the PRC Government's ongoing efforts to cut the industrial over-capacity should help to curtail emission growths. The emissions relating to steel production should fall as the industry is consolidated and excess capacity eliminated. There have also been significant improvements in energy, efficiencies especially in blast furnace operations and designs etc.

Looking ahead, the new China national ETS will take a number of years to ramp up to full sectoral coverages and to drive significant reductions. Experts have generally agreed that this ETS process should support China's overall climate actions. In addition, the new national ETS will help China to have a good system for collecting company-level carbon emission data nationally. These would help to provide a good foundation for developing and improving other future carbon policies. Cap-setting and allowance allocation practices should also support policymakers, both at the central and municipal levels, with better knowledge and confidence in developing more transparent and fairer carbon mitigation actions (China Dialogue, 2017).

Corporate carbon emissions challenges and implications

Experts have reported that some leading international corporations have been major GHG emitters. The recent Carbon Majors Report has reported that some 100 companies have been responsible for over 70 per cent of global emissions since 1988. They also reported that more than half of global industrial GHG emissions since 1988 could be traced to some 25 international companies. They forecasted that, if fossil fuels would continue to be extracted at the same rate over the next 28 years, as they were between 1988 and 2017, then the global average temperature rise could go as far as 4 degrees Celsius by 2100. This would likely result in catastrophic consequences globally including substantial species extinction and global food scarcity risks (Carbon Tracker UK, 2017).

The environmental charity CDP have also reported that major international corporations have been major GHG emitters. The Annual Carbon Majors Report for 2017 analysed data from the Carbon Majors Database which was established in 2013 by the Climate Accountability Institute. The analysis showed that carbon emissions could be directed linked to a group of major companies termed 'Carbon Majors'. The report looks at both fossil generated CO_2 and methane emissions from fossil fuel producers in the past, present and future. These provided investors and stakeholders with better understanding of the high amounts of GHG emissions from some leading companies (CDP & CAI, 2018).

These reports of the very high amounts of CO_2 and GHG emissions from the fossil energy companies have led to strong calls for these international companies to report their climate change strategies taking into context climate risks and their plans to reduce GHG emissions plus the transformation of the global energy systems. Fossil energy companies can actively contribute to the energy transition by reducing operational emissions, shifting to cleaner fossil fuels and renewables, actively deploying CCS and CCU plus other carbon-offset options including CETS, etc.

The historic agreement reached by 195 countries in Paris in 2015 demonstrated for the first time a collective determination by governments to reduce the global emissions of greenhouse gases and tackle the threat of climate change. Many governments have been introducing new energy policies to cut carbon emissions and pollutions in line with their Paris Agreement commitments. These new regulatory requirements have put pressures on businesses to put in place new carbon management strategies and to reduce their emissions.

There are many changes that companies could make to their strategies, internal processes, facilities and behaviours to reduce their GHG emissions. The management of many leading companies have recognised that they should take appropriate climate actions to 'reduce and replace', i.e. reduce GHG emissions and replace fossil fuels with clean fuels. These climate actions are within the direct control of management and are all part of good responsible business practices. However some companies and policy makers still consider climate actions could incur extra costs and would lower their short term profits. These demonstrate the importance of continued engagements by government, ethical investors and environmentalists with corporate managements internationally. Details of various investor actions and challenges to leading corporations and banks globally will be discussed further in Chapter 9, with international examples.

Looking ahead, critical strategic elements of the future carbon management strategy of international companies, especially those in the fossil energy sectors, should include carbon neutralising, carbon offsetting and complete carbon solutions. Carbon offsets should enable businesses to control their residual carbon emissions while also providing a critical source of finance for renewable energy plus emissions-reducing and resource conservation projects around the world. Tackling rising carbon levels using carbon offsetting was first adopted at the Kyoto Treaty conference in Japan in 1997. Over the past 20 years, the design, operation, and administration of carbon offsets have been improving. To date, carbon credits generated by voluntary offset schemes are subjected to rigorous industry standards. These will provide a good carbon framework with official registration systems and independent audits to ensure the reported carbon emissions reductions are real, permanent and unique. The United Nations has endorsed carbon offsetting as a valid way to reduce carbon emissions quickly and cost effectively. These should help companies to accelerate their transformation to climate carbon neutrality.

Many leading companies have recognised the importance of carbon offsets and have been actively purchasing carbon offsets. A good example is that various leading international companies have collectively invested voluntarily nearly $4.5 billion over the past decade to purchase nearly 1 billion carbon offsets from various green projects globally that have reduced or sequestered greenhouse gases. Overall, carbon offset buyers have purchased about one quarter as many offsets as the emissions they have reduced directly. Offsetting has helped to increase these companies' collective carbon mitigation impacts by some 25 per cent.

A study by Imperial College London on 59 offsetting projects showed that the purchasing of carbon credits by companies has created a host of additional benefits for companies and societies. These benefits include creating new jobs, generating new incomes, conserving local ecologies, improving access to clean water and health care, enhancing skills acquisition and promoting gender equality. The research estimated that when a business offset one tonne of its CO_2 footprint then it could generate an additional £530 (U$664) in benefits to the communities where the carbon reduction projects are based. It is important to recognise that these benefits could differ depending on the projects and their locations. This also demonstrated that with the cost of purchasing a tonne of carbon offset currently typically below €45 (US 50), the wider economic, environmental and social impacts of using carbon offsets would be considerable. The study also found that businesses investing in carbon offsets have also reported additional benefits such as enhanced brand image, motivated employees and improved market differentiation for their products or services (Imperial College London & ICROA, 2014).

In summary, carbon offsetting as part of a carbon management strategy would enable companies to cost-effectively reach the goal of carbon emissions reductions that would normally be beyond their business's economic means or technical competences. These have also provided good opportunities to accelerate the business responses to the urgent needs to reduce greenhouse gases to meet the new climate policies and clean energy targets enacted by governments in line with their Paris Agreement commitments. Research has shown that these carbon actions would also bring a wide range of environmental, economic and social benefits to the companies plus communities globally. These will also help to enhance and differentiate brand reputation for a company's products and services (BP, 2017).

Experts have advised that there are also urgent requirements to develop various innovative carbon capture and utilisation solutions, including CETS, CCU, CCS and BECCU etc. Many of these require more technical developments and cost improvements. Details of these will be discussed in more detail in Chapter 8 with international examples.

Carbon re-forestation solution developments

Many governments have been enacting new climate policies and clean energy targets in line with their Paris Agreement commitments. Experts have warned that there could be big gaps between the commitments that governments have made under the Paris Climate Agreement against the GHG emissions reductions that would be required to avoid the worst consequences of global warming. According to the UN Environment's Emissions Gap Report 2017, current pledges from governments would represent only about half of what would be required to avoid a 2°C temperature rise, and just one third of what would be required to limit warming to 1.5°C. While these climate emissions gaps are significant, UN Environment has suggested that these could still be closed in a cost-effective manner. One of the major contributors to closing the gap would be by forests. Some 6.3 gigatons (billion tons) of CO_2 emission reductions have already been reported over the past six years from forests in Brazil, Ecuador, Malaysia and Colombia alone under the UN Framework Convention on Climate Change (UNFCCC). This is equivalent to more than the annual emissions of the United States (UN Environment, 2017).

These showed that forests should be a central part of the carbon solutions for climate change and global warming globally. The UN International Panel on Climate Change, the IPCC, has suggested that if deforestation would end today and degraded forests were allowed to recover globally, then the world's tropical forests alone could reduce the current annual global emissions by 24 to 30 per cent. Hence the world's tropical forests would potentially have the capacity to contribute to about one quarter to one third of the near-term carbon solutions required to control global warming and climate change (Seymour & Busch, 2016).

Experts advise that saving forests will not only help fight climate change but will also help to protect the livelihoods of over 1.6 million people globally who have been dependent on forests for their everyday lives. In addition, enormous amounts of carbon and pollutants have been released into the atmosphere when forests were cleared and burnt by various human activities. Serious examples include extensive wood logging and drainage of carbon-rich peat swamps which have all generated significant emissions and pollutions. In particular, the burning of forests after logging has generated large amounts of carbon emissions and air pollution. A serious example is the heavy air pollution in Southeast Asia caused by the burning of forests after logging in Indonesia annually. In extreme cases, this serious air pollution has affected surrounding countries as far as Singapore and Malaysia.

8 Carbon capture and storage and utilisation innovation management

善有善报,
Shàn yǒu shàn bào,
Kind deeds will normally generate good rewards.
What goes around comes around.

Executive overview

The rising CO_2 emissions have contributed to worsening global warming and climate change impacts globally. Various improvements are being implemented including energy efficiency improvements, reducing fossil fuel consumption, increasing clean renewable energy applications etc. The application of various new innovative carbon capture technologies, including carbon capture and storage (CCS), carbon capture and utilisation (CCU) plus bio-energy carbon capture utilisation (BECCU), etc., should further reduce carbon emissions and help countries to achieve carbon neutrality globally. These carbon solutions will be discussed in more detail in this chapter, with relevant international examples.

Climate change and carbon capture storage systems developments

Climate change and global warming have been caused by rising GHG emissions. The rising CO_2 and greenhouse gas emissions have contributed to the worsening global warming and climate change impacts globally. Climate experts have warned that the various climate-induced extreme weather incidents could cause global economic losses of U$1 to 4 trillion by 2035 (BBC Environment, Carbon 'Bubble' Could Cost Global Economy Trillions, 2018).

In follow-up to the Paris Agreement, various countries have introduced different improvements to reduce GHG emissions so as to meet their Paris Agreement commitments. These improvements have included improved energy efficiency, reduced fossil fuel consumption, increased renewable and clean energy applications. The application of various innovative carbon solutions will also be very important to control global carbon emissions. These would include applying new innovative carbon capture technologies including CCS, CCU, or BECCU, etc.

Experts and leading energy companies have forecast that the continued future deployment of various fossil fuels, including oil, gas and coal, would urgently require these new innovative carbon capture technologies to be developed and deployed successfully. There is still lots of work required to develop these new carbon capture technologies, including CCS plus CCU and Bio-energy CCU (BECCU), to be both cost competitive and technically reliable so they can become part of the robust carbon solutions in future.

Albert Einstein had said that we cannot solve problems with the same approach that we have used to create them. So to mitigate the rising carbon emissions and to control global warming, the new innovative carbon capture technologies, such as CCS, CCU and BECCU etc., should have very important roles. The details of the various carbon capture innovations will be discussed in this chapter below together with relevant international examples.

CCS developments

CCS has become one of the important climate change carbon solutions. The Intergovernmental Panel on Climate Change (IPCC), has defined CCS as a process involving the separation of CO_2 from industrial and energy-related sources, then transportation to a storage location providing long-term isolation from the atmosphere. Hence CCS will typically integrate four key elements including CO_2 capture, compression of the CO_2 from a gas to a liquid or a denser gas, transportation of pressurised CO_2 from the point of capture to the storage location and then isolation from the atmosphere by storage (UN IPCC, 2005).

Experts have estimated that the contribution of CCS could be as high as 20 per cent of the required carbon emission reductions to be achieved by end 2100. Various studies globally have shown that CCS should be included as one of key emission reduction options and carbon solution for two good reasons. Firstly the cost of the overall emission reduction actions could be optimised and lowered with CCS. Secondly it is currently difficult to reduce emissions sufficiently without CCS (Haerens, 2017).

At the moment, there are around 15 large-scale CCS-projects or pilots being operated in different countries worldwide. The Global CCS Institute has estimated that these CCS projects have accounted for about 40 Mt of CO_2 every year (Global CCS Institute, 2015).

The net reduction of emissions to the atmosphere through CCS will depends on several key factors. These would include the fraction of CO_2 captured, the increased CO_2 production resulting from loss in overall efficiency of power plants or industrial processes due to the additional energy required, transport and storage requirements including potential leakages plus the fraction of CO_2 being retained in storage over the long term. A power plant with CCS could potentially reduce CO_2 emissions to the atmosphere by approximately 80–90 per cent when compared to another plant without CCS. The optimal degree of emission reductions will further depend on trade-offs between the amount of emissions reduced, the cost of carbon capture and the specific conditions of the site being deployed, such as its geology and permeability, etc.

Currently there are three main CCS technology pathways which included geological storage, enhanced oil recovery (EOR) and mineral carbonation. Storing CO_2 deep underground is generally considered currently to be the most matured CCS option. It can draw on expertise and experiences from the oil and gas sector, especially in enhanced oil recovery. A good example is that many oil companies have been applying CO_2 injections for enhanced oil recovery in various oil wells globally. It should be noted that although the individual component technologies required for CCS are well understood, the largest challenge would remain the need to successfully integrate each of these component technologies into large-scale CCS demonstration projects globally (Cuéllar-Franca & Azapagic, 2015).

Public concerns have also been playing major roles in the CCS acceptance process. Many have seen CCS as quite risky with high risk of CO_2 leakages. Some citizen environmental groups have warned of serious risks of CO_2 contaminations of ground water and alteration of groundwater chemistry, especially in the USA and UK. It would be important to provide adequate and unbiased information to stakeholders to mitigate these concerns about CCS geological storage, potential leakages and behaviour of CO_2 in the reservoir (Global CCS Institute, 2015).

The European Union has included CCS in its EU Climate and Energy Package as a viable option to achieve the long-term EU emissions reduction goals. CCS has been accepted as one of the few potential viable options to reduce unavoidable emissions from industrial processes, such as the production of steel, on a sufficiently large scale. A legal framework, the EU 'CCS Directive', was set up to ensure safety as well as to minimise risks and negative effects. Even though the European Union is providing support, currently there have been no large-scale CCS projects operational in the European Union though some are being considered (European Commission, 2012).

In the UK, a pioneering CCS competition was launched for a possible one billion GBP investment in a potential new CCS demonstration project. However the sponsors decided to exit the competition in November 2015, just six months before it was due to be awarded. The argument was that CCS did not fit into the areas of spending that would offer the best economic returns whilst delivering on the commitments to invest 100 billion GBP in infrastructures. However, the UK has been reconsidering different CCS projects for potential future investments (UK Parliament, 2016).

CCS technology developments overviews

There are currently three key CCS technologies globally. These will be discussed in more detail below, together with relevant international case examples.

The CCS geological storage technology would consist of capturing CO_2 from point sources at power plants or industrial installations, compressing it, transporting it and injecting it into suitable geological formations. CO_2 would either be stored as a compressed gas, a liquid or in a supercritical state in various designated locations, including geological and ocean storage. In the geological storage applications, the CO_2 would normally be stored at depths of 800–1000 metres in geological formations such as depleted oil and gas reservoirs, deep saline aquifers and coal bed formations. In the case of ocean storage, the captured CO_2 would normally be injected directly into the deep ocean to depths greater than 1000 metres. In contrast to geological storage, ocean storage still has not been tested on a large scale, even though it has been studied for over 25 years. There have been concerns about potential biological impacts, high costs, impermanence of ocean storage plus public acceptance concerns. A good example of CCS geological storage that has been demonstrated successfully is in Norway at the Sleipner gas field, which has been operating since 1996.

Enhanced oil recovery is a CCS technology which has been actively applied in the oil and gas sector. These have helped to produce more oil from various discovered oil fields globally after it has already gone through the primary and secondary oil production phases. After the primary and secondary oil production phases there would normally still be about 50–70 per cent of oil remaining in the reservoir. The operators could then choose to produce more oil in a tertiary oil recovery phase by injecting heat, chemicals and/or gases, including CO_2, into the oilfields. These techniques would be part of the overall enhanced oil recovery (EOR) package. CO_2 flooding or CO_2-EOR has been one of the most proven EOR methods globally. When almost pure (at least 95 per cent) CO_2 is being injected into a depleted reservoir, it will make the oil swell and become lighter. This results in the oil detaching away from the rock surfaces. The oil is then able to flow more freely within the reservoir, so that it can be collected.

These enhanced oil techniques, including CO_2-EOR, have led to higher oil productions and recoveries globally. Best EOR practices have generally helped to recover an extra 5–15 per cent of the original oil in place (OOIP). Usually the CO_2 and water would be separated from the oil when these reach the surface. Then they could be recovered and re-injected for further EOR.

Similarly, CO_2 could be used to extract more natural gas from coal deposits which would normally be un-mineable, by the Enhanced Coalbed Methane Recovery (ECBM) technology. For difficult gas fields, the Enhanced Gas Recovery (EGR) technology could be applied (Cuéllar-Franca & Azapagic, 2015).

Good examples of CO_2-EOR can be found in America. The use of CO_2 injection and flooding to enhance oil production from oil wells has been widespread in the USA and Canada. The first large-scale testing of CO_2-EOR took place in the 1970's in the high Permian Base in Texas and New-Mexico in USA. The injected CO_2 would normally originate from three different sources, firstly from natural hydrocarbon gas reservoirs which normally contain CO_2 as an impurity; secondly from natural CO_2 reservoirs; and thirdly from industrial or anthropogenic sources. In all three cases, the CO_2 rich gas would need processing first in order to bring the CO_2 concentration to the right levels of 90–98 per cent. The main source of CO_2 used today would be naturally occurring CO_2, because of its lower cost and wider availability. Currently, about 45 Mt of naturally stored CO_2 is being used annually for EOR or EGR (Cuéllar-Franca & Azapagic, 2015).

Even though the monetary benefits of CO_2 enhanced oil recovery applications would not occur as quickly as in gas and oil drilling and exploration, the long term benefits of CO_2-EOR could be substantial. A good example is the first CO_2-EOR flooding projects in the Permian Base in Texas, USA in the 1970s. These are still under operation today and producing nearly one million barrels of oil per year. Forty years after the reservoir was denoted as depleted, CO_2-EOR has helped to create more oil and gas productions. These have helped to generate more revenues, taxes and employment plus CO_2 storage at the same time. These large economic and social benefits of enhanced oil recovery would offset some of the capture and storage costs, making the CCS option more cost-effective.

Mineral carbonation is another CCS technology option that has been applied in different countries. This involves reacting the CO_2 with metal oxides, resulting in the formation of carbonates with heat releases. This is a natural process which will permanently fixate CO_2 in a solid mineral phase. There is a difference between in situ and ex situ carbonation. The former is similar to geological storage, since CO_2 is injected into silicate-rich geological formations or alkaline aquifers, resulting in carbonates. The latter involves carbonates, natural minerals or industrial residues, such as slag from steel production or fly ash, and would require additional energy inputs. A good example is the recent CarbFix Project in Iceland. In this project, CO_2 emissions from a geothermal power plant were injected into basaltic rocks of 400 and 800 metres deep. After two years, about 95 per cent of the injected CO_2 had been fixated. These had reacted with the volcanic rock to form carbonate minerals at rates much faster than predicted and at a relatively low cost (17 USD/t CO_2). These results demonstrated that CCS carbonation could be a safe long-term carbon storage solution for anthropogenic CO_2 emissions (Cuéllar-Franca & Azapagic, 2015).

CCU developments

CCU processes have been developing recently as one of the future key climate change carbon solutions. CCU is different from carbon capture and storage CCS in its final treatment of the captured CO_2. Instead of transporting the CO_2 to a particular location for long-term storage, CO_2 would be used as a raw material and transformed into value-added products with new innovative technologies (Cuéllar-Franca & Azapagic, 2015).

CO_2 emissions could be reduced via two major routes. Firstly, directly through consuming CO_2 and thereby avoiding its release in the atmosphere. Secondly, converting or substituting CO_2 into useful products or services. New innovative carbon capture and utilisation CCU technologies aim to reduce CO_2 emissions whilst creating useful commercial products. These could then lead to new clean business opportunities for green product developments.

At present, the overall CO_2 emissions reduction potentials would be limited by the available market potentials for CO_2-based products. One scientific fact is CO_2 is a relatively inert stable molecule. So most of the current industrial uses of CO_2 are highly energy intensive. Some good examples of the current industrial uses of CO_2 would include the production of urea with some 70 Mt CO_2 per year, inorganic carbonates and pigments with some 30 Mt CO_2 per year, methanol with some 6 Mt CO_2 per year, salicylic acid with some 20 kt CO_2 per year and propylene carbonate with some thousands of tonnes CO_2 per year. Looking ahead these markets are likely to grow significantly as new CCU innovations and technologies become available. A good example is the new MTO/MTP technologies and process innovations which have led to many new plants being built and planned in China and globally.

Experts have estimated that if the currently known CCU processes were to be deployed in the most efficient way then they could potentially use up about 300 million tonnes of CO_2 per year. These should help indirectly to reduce CO_2 emissions by around one 1 Gt per year, which would be about 5 per cent of the total net CO_2 emissions globally.

The first CCU option is the direct utilisation of CO_2. Pure CO_2 is currently used directly in various applications. In the food industry, CO_2 is used to carbonate beverages, soft drinks and in food processing, preservation and packaging. Furthermore, CO can be used to decaffeinate coffee by bathing the steamed coffee beans in compressed CO_2, which would then remove the caffeine without eliminating the flavour. For the food and beverages markets, high purity CO_2 streams will be required.

Other CO_2 direct applications have included the provision of CO_2 to greenhouses to maximise plant growth rates, the use of CO_2 as a refrigerant gas in large industrial air conditioning and refrigeration systems plus in fire extinguishers and dry fabric cleaning etc. In the pharmaceutical industry, CO_2 have been used as a respiratory stimulant or as an intermediate in the synthesis of drugs. Most of these applications are restricted to processes which would produce CO_2 offgas streams of high purity, such as ammonia production. Moreover, the CO_2 sequestration capabilities of these markets are rather low. Good examples are over 10 million tonnes Mt of CO_2 have been applied in the food and beverage industry plus 6 Mt CO_2 as an industrial gas.

Another interesting CCU possibility is the conversion of CO_2 to clean fuels. These would involve the hydrogenation of CO_2 via new innovative processes. A good example is CO_2 and hydrogen (H2) being brought together via new innovative catalytic processes to produce large volumes of chemicals such as methanol, dimethyl ether (DME) and ethanol, which can then be used as clean fuels. The main concern for this process is the availability of hydrogen. Hydrogen could be produced through water electrolysis but this will require a lot of energy thus increasing the cost significantly. The averaged manufacturing cost of hydrogen, produced by alkaline water electrolysis with wind powered electricity, would correspond to over 6–7 $/kg depending on the clean electricity sources. Ethanol also has several advantages over methanol due to its safer handling and better compatibility to gasoline. Other hydrocarbon fuels could also be produced using hydrogenation reactions. However these would require high energy consumptions resulting in higher costs.

CO_2 could be utilised as a feedstock to produce useful chemicals. The conversion of CO_2 to urea has been significant but conversion to other chemicals has been limited to a few applications on a modest scale. Urea synthesis has accounted for approximately 130 Mt CO_2 per year. Aside from that, only a marginal amount of CO_2 has been utilised to produce various speciality chemicals (Cuéllar-Franca & Azapagic, 2015).

An exciting new CCU chemicals development has been the new MTO/MTP process for the conversion of CO_2 to chemicals. These would normally involve the catalytic hydrogenation of CO_2 together with synthesis gas reactions. A good example is CO_2 and hydrogen (H2) being brought together via new innovative catalytic processes to produce large volumes of chemical intermediates such as ethylene, propylene, methanol, dimethyl ether (DME). These can then be converted to useful chemical products via traditional chemical processes. The main constraints would the availability of suitable large quantities of purified CO_2 required for economic operations of these big complex industrial chemical plants.

Photo- and electrochemical/catalytic conversion of CO_2 is another interesting new possibility. Solar energy can be used to directly or indirectly reduce CO_2 via biological routes. This biological photochemical reduction of CO_2 would be based on reproducing nature's photosynthesis process. Sunlight, water and CO_2 would combine over new innovative catalysts to reduce CO_2 to CO and then convert it into other interesting organic compounds. Potential challenges include the efficiency of the catalyst necessary for the reaction plus the cost of the materials used for synthesis. A lot of advances have been made in new catalysts in recent years and there may be interesting breakthroughs in future.

There have been several companies which have successfully operated CCU plants with clean renewable energy to chemically produce green fuels from CO_2. The current market leader has been Carbon Recycling International (CRI) which is an Icelandic company. It has been producing methanol based on CO_2 and hydrogen with renewable energy. It has developed an innovative process to use the abundant geothermal energy available in Iceland to split water into hydrogen and oxygen. CO_2 and hydrogen are then reacted to produce green methanol which have been sold for blending with gasoline to produce biodiesel. The CRI-plant has been shown to generate 90 per cent less CO_2 emissions, as compared to the similar fossil based productions. The annual production of green methanol has amounted to about 5 million litres.

The conversion of CO_2 into useful materials via mineral carbonation could be another potential CCU option. CO_2 can be used to react with various industrial residues, such as steel and blast furnace slags, cement kiln dust and waste cement, fly ashes, municipal waste incineration ash, mining wastes and asbestos etc. These have good CO_2 fixation potentials and can turn CO_2 into useful commercial products after reactions. A good example is the production of cement and building materials out of stainless steel slag. This is currently being done by Carbstone Innovation NV which is a Belgium-based company. The current capacity of these applications is quite small, when compared to the amount of CO_2 requiring treatment globally. These technologies need further developments to demonstrate their commercial viability on larger industrial scales.

Bio-energy CCU (BECCU) developments

Bioenergy with carbon capture storage and utilisation (BECCU) processes are being developed as potential new carbon technologies. The challenges associated with each BECCS technology are large and complex. Biological carbon mitigation involves CO_2 uptake by living organisms, which can be via photosynthetic or electro-synthetic processes. Photosynthetic microorganisms will use solar energy to convert CO_2 into organic carbon with a significant amount of biomass. Experts estimated that these organisms have already been converting around 100 Gt of carbon into biomass annually in nature. These have involved some highly sophisticated natural mechanisms for carbon fixation and utilisation which are being studied. These bioprocesses could potentially lead to new disruptive carbon technologies which will need more developments.

Similarly, micro-algae could be used to convert CO_2 into clean biofuels. There has been ongoing research on optimising micro-algae's higher photosynthetic efficiency to convert CO_2 into valuable biofuel and non-fuel co-products. The photosynthetic efficiency of micro-algae could be in the range of 3–8 per cent, which is much higher than that of terrestrial plants, which is only 0.5 per cent on average. The microalgae could help to convert CO_2 to biofuels, biodiesel, fermentative bioethanol and biobutanol. The cultivation of algae could take place in open-pond systems or in new innovative closed bioreactors, with good supplies of water, nutrients and CO_2 plus continuous mixing. After the photosynthetic biomass production, the biomass would need to be separated and processed into biofuels suitable for market supplies. A potential industrial application being developed is to use micro-algae to biologically fix CO_2 from flue gas emissions from thermal power stations. The large CO_2 emission requiring treatment could make the integrated microalgal bioreactor and biorefinery potentially more economic.

Lastly CO_2 could also be sequestered by anaerobic bacteria for metabolism under specific conditions. This process is enabled by the catalytic activity of certain enzymes that are present in anaerobic CO_2-sequestering organisms (e.g. Archaea). The end products of this anaerobic fermentation process would be chemicals, such as alcohols and biogas. These could be used as biofuels or as bio-feedstocks for conversion into green products. For these metabolism processes, the operational parameters such as pressure, temperature, hydraulic retention time, etc., are of extreme importance.

A commercial application of this innovative anaerobic technology has been applied by LanzaTech, a company founded in New-Zealand. They have discovered suitable microbes which would treat waste gas from power plant, steel mills and so on, in the LanzaTech Biological Fermenter. The CO_2 would be converted by the microbes in the reactor into biofuels, including ethanol and byproducts. These would then be separated in a downstream hybrid separator into sellable biofuels and byproducts for sale to the local markets. Currently, it has been successfully operating two plants at steel mills in China, where commercial operation has begun in 2015. Ethanol produced this way, would lower the equivalent carbon footprint by 60–80 per cent (LanzaTech, 2019).

In summary BECCU processes could help to convert CO_2 into useful biofuels and byproducts. Further process developments are required including detailed evaluation of lifecycle CO_2 balances and biodiversity impacts. Good governance and financial incentives are also required to stimulate BECCS industrial application developments (Imperial College London, BECCS Deployment, 2019).

China carbon solutions developments case study

China has been investing heavily in new carbon capture and utilisation (CCU) technologies. A key driver is that new CCU innovations could help China to capture the large amounts of CO_2 generated by fossil fuels and convert it into useful clean biofuel or green chemical feedstocks. These would help to reduce the current high oil and gas imports into China plus improve energy security. Alternatively they could store the CO_2 in suitable underground rock formations with CCS. These should help China to reduce their GHG emissions and meet their Paris Agreement commitments.

China has been undergoing major energy transformations recently with government supports. They have been reducing fossil fuels consumptions, especially coal and oil. There have been major investments into clean renewable energy facilities. Despite its enormous progress in renewable energy, China has still been using a lot of coal for power generation and chemicals manufacturing. China currently has more than 900GW of coal-fired power stations which have been generating about two-thirds of China's electricity and more than 80 per cent of its CO_2 emissions. Although coal consumptions in China have started to decline for the first time in 2015, additional new coal power generation capacities, up to 200 GW, are being planned to be brought on stream in the next few years. Energy experts have predicted that coal would still be producing well over half of China's energy by 2030. Although efforts have been made to move away from coal power, it would take a long time for China to wean itself off coal completely.

Nevertheless, there could be a gradual shift in how the coal could be used with new cleaner approaches. The coal could be used with new processes to produce syngas as a clean feedstock for the chemical industry. Whilst this is not a complete removal of fossil carbon, it would put it to a better use from an environmental point of view. Coal would typically be treated with oxygen and steam in advanced gasifiers to produce syngas, which is a mixture of hydrogen and carbon monoxide. The syngas, after treatment, would then go through the Fischer–Tropsch FT process to produce clean liquid hydrocarbon fuels and to produce chemicals like methanol, ethylene, propylene, etc. via the new MTO/MTP processes. Looking ahead, these new carbon processes could help to reduce CO_2 emissions on a large scale in China. There has been ongoing research in leading Chinese Universities and Research Institutes with generous funding support.

China has various CCU projects in the early stages of development. Experts predicted that China would become a key global player in CCU over the next two decades. A good example is that in 2017, construction began on the Yanchang Integrated CCU Demonstration facility in Yulin City. This CCU pilot have been designed to capture about 400,000 tonnes of CO_2 per year from two coal-to-chemicals plants nearby. The CO_2 would then be pumped into the nearby Qiaojiawa oilfields for enhanced oil recovery (Global CCS Institute, Yanchang Petroleum Report 1, 2015).

The typical CCS and CCU capture processes would currently use solvents such as mono-ethanolamine to react with CO_2 to form C–N bonds. These would subsequently break apart at 100–120 degrees Celsius to release purified CO_2 that would be suitable for storage or utilisation. One of the big technical challenges has been that the temperature swings involved in capturing, separating and releasing CO_2 in this way would make it a very expensive process. The costs have been estimated to be around \$50–100 per tonne of CO_2 and consuming 25–40 per cent of the energy generated by the plant. This would effectively double the cost of the electricity it would generate.

So researchers have been investigating a range of other options to capture CO_2 with lower regeneration costs. These include new porous materials that could physically absorb CO_2 and then release it under much milder conditions. A good example is that a team at SunYat-Sen University in

Guangzhou has developed a series of porous poly-amine materials that would readily adsorb CO_2 at room temperature, and then desorb the gas at 90 degrees Celsius.

Other researchers have been investigating metal–organic frameworks (MOFs) for CO_2 absorption. These are highly porous materials made from metal-based nodes held together by organic struts or linkers. MOF has good material advantages as their structures and chemical activities are highly tunable. It should be possible to design materials which would bind CO_2 sufficiently well to separate it from a gas stream, but would also release the CO_2 without extreme heating. Using new design approaches, the Sun Yat-Sen University team have found that a cobalt-azolate MOF known as MAF-X27ox has the highest uptake of CO_2 by volume of any MOF. It would also release it rapidly at a relatively mild 85 degrees Celsius (ACS, 2018).

Once CO_2 has been captured, storing it underground should be a last resort. A much better option would be to transform the CO_2 into useful chemicals. Some of China's leading R&D work in this area is being undertaken at Dalian. The Dalian Institute of Chemical Physics DICP, is also home to the Dalian National Laboratory for Clean Energy (DNL), which opened in 2011. It was China's first national laboratory in the field of clean energy research.

On CCU, a DNL research team unveiled in October 2017 they have developed a novel zinc oxide–zirconium oxide catalyst that would convert hydrogen and CO_2 to methanol. Previous catalysts for this reaction had suffered from poor methanol selectivity and produced a messy mixture of products or deactivated rapidly. In contrast, the bi-metallic catalyst remained stable for more than 500 hours with good selectivity producing over 90 per cent of the end product as methanol. One reason for the success was that the two metals have different functions. Zinc would help to bind H2, whilst neighboring zirconium would bind and activate CO_2.

A DNL team has also developed a catalyst that could turn CO_2 and hydrogen into oil products with gasoline-range hydrocarbons. The catalyst was composed of iron oxide nanoparticles. Their surfaces have been peppered with sodium atoms, which were loaded into a zeolite. This effectively provided three different catalytic sites. The CO_2 is first reduced to CO, then hydrogenated to produce olefins, and finally coupled together inside the zeolite.

In summary, the various new carbon process innovations in China could help to significantly reduce its carbon emissions whilst helping to convert CO_2 into useful biofuel plus chemical products. More developments would be required but good progress is being made to date.

Carbon solutions in energy and chemical sectors overviews

Globally the fossil energy industries have been accountable for about 60 per cent of the global CO_2 emissions. These have been generated mainly from fossil fuels combustions, including coal, oil or natural gas. These fossil fuels have been providing over 80 per cent of the global energy supply over last few decades. It is interesting to note that in producing the same amount of energy, coal is almost twice as polluting as natural gas whilst emissions from oil products would lie in between gas and coal (IEA, World CO_2 Emissions from Fuel Combustion, 2015).

In contrast with the steel sector, the energy sector has a plethora of possibilities to generate power that do not lead to the emission of CO_2. These would include applying renewable energies such as wind, solar, hydro, geothermal and nuclear power. These are all proven technologies which would generate energy without the release of greenhouse gases. In addition, there have been major innovations and serious costs reductions in all the renewable technologies in recent years. Experts have forecast that clean renewable energy should become cost competitive with fossil fuels by 2020–2025 period, with the ongoing rate of innovation and cost reductions.

Most leading energy companies globally have been slowly transitioning their corporate strategy towards including clean renewable energy into their business portfolios in addition to fossil fuels. Most energy companies have considered these strategic transformations as essential elements of their climate change and sustainability strategies. The pace of their energy transformations would be heavily influenced by the prices of oil, fossil fuels, plus regulatory pressures etc. A good example is that experts have forecast that if both the oil and EUA prices would go up significantly in future, then the rate of transition from fossil to renewable energy would be speeded up even more.

Global investments in renewable energy has been growing swiftly. A good example is that annual investments of some 200 billion USD per year have been made in renewables in recent years globally. As the investments in renewables have gone up, the costs have also been reduced with various innovations. A good example is that the cost of a typical photovoltaic (PV) rooftop system in Europe has decreased by 90 per cent in the last 25 years, with new innovations and improvements.

Energy companies have also been looking at various carbon technologies from different perspectives versus iron and steel companies. Looking ahead, some fossil energy companies have predicted that in the longer term future their fossil power plants would have to be shut down entirely and be converted to renewables. The situation of the energy sector is different from the steel and cement industry. For the steel and cement sector, they have to apply CCS and CCU to reduce CO_2 emissions into the atmosphere. A good example is that some steel plants have been investigating using slag to fixate CO_2 emissions.

Most of the CCS and CCU developments in the energy sector have focused on major power plants as they are ubiquitous and would provide the largest source of CO_2. One of the most interesting new technological developments in CCU would be to convert CO_2 to useful biofuels and chemicals. The global market demands for clean fuels and green chemicals are of considerable size. There are still many technical and commercial hurdles on the CO_2-to-fuels/chemicals pathways. These include high costs in both high operational expenditure (OPEX) and capital expenditure (CAPEX). The CO_2-based synthetic fuels and chemicals would have to compete with other fossil-based fuels and chemicals. The regulatory framework surrounding CCU is also unclear. A good example is that it has been unclear whether converting CO_2 to fuels or chemical products would mean that no carbon allowances would have to be paid in the EU under the EU-ETS.

There are also interesting new business cooperation opportunities in CCS and CCU applications between energy, cement and steel companies. A good example is that some leading energy companies have started to offer new integrated energy and carbon solutions services which include CCS and CCU technologies for their own uses as well as a service to their clients. An interesting potential option is that a cement company, which generates CO_2 as a by-product of its cement production, could consider forming a win-win partnership with an energy company to process its CO_2 emissions. The energy company could then convert the CO_2 to synthetic fuels and chemicals via innovative new technologies. Then the partners could sell these valuable products to the markets whilst also reducing their CO_2 emissions.

The chemical industry is in an interesting position with regard to CCU technologies. The chemical industry globally has been responsible for generating close to 20 per cent of industrial CO_2 emissions globally. Looking ahead, there may be good business opportunities for the chemicals companies to provide valuable knowhows for the new CO_2 chemical conversion processes for themselves and their clients globally. Hence the chemical companies are on both the supply and demand sides of CCS and CCU technologies.

Many chemical companies are more interested in the supply of carbon monoxide (CO) instead of CO_2. The reason is that CO is thermodynamically more efficient then CO_2 and CO is already used as a raw material, as part of syngas, for multiple applications in chemical companies. The costs of producing chemicals out of CO is however still much higher than conventional fossil oil or gas processes. The chemical products made from CO must compete with their fossil-based equivalents in the global markets. Several drivers could help to offset these economic hurdles. Firstly, an increase in the oil price could reduce the differences in production costs between fossil-based and CO-based chemicals. A second driver would be to give official recognition that CO-based chemical products are more environmentally friendly than their fossil-based equivalents. This could be done by specific policy support or by means of sustainability labels. This would then help the growth of a new market for green chemicals, in which customers would pay a premium price for these green products. In addition, these should help to create better corporate branding and higher competitive advantages plus attract more customer preferences.

There should also be interesting win-win business cooperation opportunities by chemical companies with steel companies in the treatment and reduction of carbon. A good example of win-win cooperation between chemical and steel companies is in steel mill gas. Normally steel mill gases will contain large amounts of CO, aside from CO_2. When this CO is emitted into the atmosphere, it would react with oxygen to form CO_2 in just a matter of seconds. By extracting the CO from steel mill flue gas, CO_2 emissions would be indirectly reduced, because the CO would not then be reacted to CO_2. The extracted CO can be treated and then used to produce useful chemical products. So this new integrated CO process should help to reduce GHG emissions plus improve useful byproduct manufacturing via new CCU technologies. More work would have to be done to further develop these promising new technologies so as to reduce CO_2 emissions and minimise global warming.

9 Climate finance and climate risk management

国以民为本
Guó yǐ mín wéi běn
The foundation of a country is provided by its people.
People are a country's roots

Executive overview

Globally there are urgent needs to control climate changes and manage climate risks. Climate finance and green investments have been growing in recent years. These should help to improve sustainability and environments globally. Climate change risks could seriously undermine the financial performances and the bottom lines of leading banks and companies globally. New climate risk financial disclosure and governance requirements are being introduced by various key governments and stock markets globally. These would compel all the leading banks and companies internationally to improve their corporate governance and financial reporting. Details of these will be discussed in this chapter, with international examples.

Climate change and climate finance developments

Climate change and global warming have generated serious negative impacts on global economies. These have led to many climate-induced extreme weather events such as hurricanes, typhoons, heavy rainfalls, flooding and droughts. These have caused huge economic damages with large financial costs, in various cities globally. There are urgent needs for governments and companies to act quickly so as to control the rise of global warming and minimise the impacts of climate change. A good evidence of serious climate damage is that an Aviva-sponsored Economist Intelligence Unit report has estimated that up to US$43 trillion of assets could be lost by 2100 as a result of extreme weather events induced by climate change (Economist Intelligence Unit, 2015).

The signing of the Paris Agreement on Climate Change in December 2015 marked a key milestone in climate change and green finance. In combination with the 2015 Sustainable Development Goals, these demonstrated for the first time the worldwide commitment to fight global warming and climate change so as to achieve a global low carbon future. These also signaled a universal shift towards a less carbon intensive and more climate-resilient economy. The Paris Agreement stimulated leading countries to joint climate actions and promoted green investments for climate improvements. These major new green investments should help the global transition to low carbon economy which should support sustainable economic developments globally. The UN Sustainable Development Goals (SDGs) have set out specific goals to be achieved in the next 15 years to end poverty, protect the global environment and promote sustainable economic developments for all. It also implied a significant rethink of the world's current economic development models to support future sustainable economic developments globally (UN, Sustainable Development Goals (SDGs), 2015).

The Paris Agreement, in combination with the SDGs, has demonstrated the worldwide commitment by leading countries to work jointly together to manage climate change and improve sustainability globally. They have aspired to work together to achieve a low carbon future for the world to transform their economies so that these will become less carbon intensive, more climate-resilient and with improved sustainability (UNFCCC, Paris Agreement, 2016).

The New Climate Economy Report has estimated that around US$90 trillion of new green investments would be required between now and 2030 so as to achieve the global sustainable development objectives. After the signing of the Paris Agreement, many leading countries and key players globally have been implementing urgent measures to control climate changes and manage climate risks. Climate finance and investments have been growing strongly in recent years. There are plans for further new green finance investments to reduce emissions and lower global warming. These should help to improve sustainability and environments globally. There have also been urgent needs to improve the governance and robustness of leading banks and companies globally so as to better manage the negative effects of climate change and the associated climate risks (Global Commission on the Economy and Climate, 2014).

Global climate finance and green investment momentum have been building up in the last few years. Economists have reported that over US$3.5 trillion of new private climate finance and green investments have been mobilised to date. The global green bond markets have been growing fast and reached US$155.4 billion new issuance in 2017. This was almost double the US$81.6 billion of green bond issued in 2016. The globally managed assets in the environmental and sustainability sectors have also increased by 25 per cent from 2014 to 2016, with further growths expected in future. The annual global investments in clean energy have grown seven-fold, from US$47 billion in 2004 to over US$335 billion in 2017 (UNFCCC, Biennial Assessment and Overview of Climate, 2016).

Many developing economies, especially China, India and in Latin America, have been growing fast. They will be contributing to greater shares of the global Gross Domestic Product (GDP) and future world trade growths. The demands for green finance and green investments are also great in these fast growing developing economies. In their transition to new low carbon economy, these would generate significant new trade opportunities globally. A good example is that in China, it is estimated their new 'ecological civilisation' transformation would require new climate investments of US$470 billion to US$630 billion in the period 2015 to 2030. Experts have forecasted that the PRC Central Government would only provide some 15 per cent government seed green investments. At least 85 per cent of the planned new investments will have to come from the private sector in China.

These important strategic drivers have led to a significant growth in climate investments globally in recent years and their growth momentums are likely to continue in near future. These new climate investments should help to contribute to global climate change, environment and sustainability improvements.

Climate risks financial reporting and governance improvements

Globally there have been rising concerns from governments and regulators about the implications of climate risks on leading banks and companies globally plus how well prepared they are to handle these. Climate risks could pose serious environmental and reputational threats which could seriously affect the business performance and financial results of leading banks and companies globally. There were serious concerns by the G20 Finance Ministers and key Central Bank Governors that the financial implications of climate change have not been adequately disclosed by leading corporates globally to the finance markets and their investors. Whilst corporates might have taken into account the physical climate risks, there have been concerns that the climate transitional risks have been ignored. Insufficient disclosures would also undermine the international capital markets from making good asset allocation and risk pricing decisions. All these could contribute to serious market and financial instabilities globally, which would have serious negative impacts on global economies.

There have also been serious rising shareholder demands and pressures on the leading companies and banks to improve their climate risk and financial reporting. A good shareholder action example was that in Australia in 2017, the shareholders of the Commonwealth Bank successfully sued the bank for failure to adequately disclose in their Annual Report the business risks induced by climate change. The shareholders cited a wide range of potential climate risks, especially the investments by the Bank in the extractive industry and the housing industry and markets in Australia. The shareholders have launched their successful legal challenges coinciding with the serious warnings by the Australian financial regulator that climate change risks could create serious material risks and damages to the financial systems of Australia.

There have also been significant growths of the ethical impact investor movement globally. Recently some 100 major ethical impact global investors, who are collectively holding some US$ 1 trillion of financial assets and investments globally, have jointly sent an important letter to each of the top 60 largest banks in the world. In their joint shareholder letters, they have requested the board and management of each of these 60 top banks to clearly and properly inform their shareholders about how their investments in these banks would be affected by climate change risks.

In the face of all these intense international, government and shareholder pressures, leading banks and companies globally have to take the new requirements on climate related financial disclosures very seriously. Their board and top management have to put in place appropriate new management systems and processes to compile with the new disclosure and reporting requirements. They should prepare the appropriate climate strategies and risks reports which would meet the new international requirements. Otherwise they would be opening themselves to serious challenges and even lawsuits from the regulators, investors and media.

G20 Task Force on Climate-related Financial Disclosures (TCFD) recommendations

The G20 Finance Ministers and Central Bank Governors have a lot of concerns that leading companies and banks internationally have not been adequately disclosing the financial implications of climate change and climate risks to the international markets and investors. They felt that whilst corporates might have taken into account the physical climate risks, the transitional climate risks have often been ignored. They consider that poor disclosures would hinder the international capital markets from making well-informed credit rating plus asset allocation and risk pricing decisions. These could lead to serious financial and market instability problems, which could lead to another global financial crisis.

The G20 Finance Ministers agreed that the first step towards better management of climate change and its risks will be accurate measurement and reporting. They appointed Mark Carney, the Financial Stability Board (FSB) Chair and the Bank of England Governor, to create the G20 Task Force on Climate-related Financial Disclosures (TCFD) in December 2015. The TCFD was chaired by Michael Bloomberg and comprised 32 financial leaders across the world. They have undertaken extensive engagements, surveys and reviews in developing their recommendations and report.

In June 2017, the G20 Taskforce on Climate Financial Disclosure TCFD published its final report and recommendations, which contained detailed recommendations on voluntary climate-related financial disclosures. The TCFD report recommended leading banks and companies should make their future climate disclosures in line with a new framework underpinned by seven key principles. Their recommendations sought to balance the global need to raise the bar for existing climate financial disclosure standards together with the desire to achieve widespread adoption by leading corporates globally. They have provided new guidelines on how leading corporates globally should use their mainstream, and publicly available, financial reporting instruments to report on their climate change risks and strategy plus opportunities that they would face in the short-, medium- and long-term. In general, the TCFD recommendations represented very carefully considered, open and transparent new approaches for climate-related financial disclosures by leading banks and companies. The TCFD team has also identified some key industrial sectors globally that could suffer higher climate risks. These sectors included the energy sector, transportation, construction, agriculture, food and forestry (G20 TCFD, 2017).

In the April 2017 meeting of the G20 Finance Ministers and Central Bank Governors, they generally accepted the TCFD recommendations and supported their voluntary adoption by companies and banks globally. Looking ahead, some key G20 countries would be introducing new regulations to enforce the TCFD compliance. Other countries might prefer softer approaches such as recommended guidance for voluntary adoption. It is generally believed that these new TCFD recommendations would become important new international requirements that leading companies and banks globally would have to comply with in future. Otherwise they could face serious challenges and even lawsuits from the regulators, investors, shareholders, stakeholders and media.

Recognising that these new TCFD approaches would evolve over time, the Task Force developed a set of seven key principles to underpin the new disclosures that companies should prepare. The report recommended leading companies and banks globally to report on four key corporate areas covering climate governance, climate strategy, climate risks and management targets.

On climate governance, it has been recommended that the boards of leading banks and companies should disclose details of their new governance processes and systems for assessing and managing climate risks and opportunities. These should include the governance systems via which the boards would be overseeing climate governance plus the associated governance management processes. These should include how frequently the board would be informed on progress and how they would review these with their top management. In addition, the boards should disclose how they would be assessing the actual governance outcomes against major business plans and risks. In addition, the boards should define what would be the top management's roles and accountabilities in addressing these climate-related risks and opportunities.

On climate strategy, the boards and senior management would be required to report on the various climate strategies and scenarios that they have considered. These should include what would be the actual and potential impacts of climate changes on their business performance and financial results. They should also consider climate impacts on their business organisation, strategy and financial planning.

On climate risks, the senior management should give details on their climate risk assessments and risk management systems. These should include how they actually evaluate, identify, assess and manage climate risks plus opportunities. In addition, they should disclose their risk management systems and action plans as well as the risk mitigation plans and actions they have formulated to minimise the impacts of these climate risks on their businesses.

On climate metrics and targets, the board and senior management should give details of the appropriate climate metrics and targets that have been established for their management and businesses. These new metrics and targets should ensure that management and staff are managing climate risks effectively.

It is important for international banks and companies globally to recognise that climate change impacts and risks on their businesses could also be influenced by new government policies and changes in the law. A good example is the changes in the energy policies for fossil fuel and renewable energies being introduced by various key countries globally as part of their Paris Agreement commitments. There are also rising risks of litigation by regulators plus challenges by ethical investors and shareholders globally.

In addition, specific physical climate risks could affect various businesses differently. Typical good examples would include the availability of suitable raw materials and clean renewable energy, scarcity of water plus how to source these on a sustainable basis. All of these physical risks could seriously affect a company's revenue, expenditure, assets, liabilities, capital and their financing. In addition, it is important for management to recognise that there could also be new interesting low carbon business opportunities created by climate change. These could include new markets, new employment, new green financing, new green investments to improve resource efficiencies and energy efficiencies, etc.

Global sustainability and governance developments

Globally there have been serious concerns from various governments about the implications of the various serious climate risks on leading banks and companies plus how prepared are they to handle these. International capital markets would also all require high quality, accurate and timely data from their leading banks and companies so they could perform their capital market function efficiently. Investors and fund managers would also need better information to make informed investment decisions about the impact of climate risks on their business operations and future capital investments (GFI, 2017).

The G20 private-sector-led TCFD has provided an internationally agreed framework through which exposure to climate risks could be assessed, managed and reported. The G20 Finance Ministers, in their April 2017 meeting of the G20 Finance Ministers and Central Bank Governors, have generally accepted the TCFD recommendations and supported their voluntary adoption by companies and banks globally.

It is generally believed that the G20 TCFD recommendations will become an important new international requirement. Leading companies and banks globally will have to do their best to meet and comply with these requirements. Many key governments, such as UK, France, Belgium and Germany, have supported the introduction of the TCFD framework in their countries. Some countries have favored developing new regulations in future to enforce the TCFD recommendations. Some other countries might prefer softer approaches such as recommended guidelines for voluntary adoption. Looking ahead, different governments would follow different paths for compliance. The implementation pathways adopted by some key governments, including UK, China, EU and Hong Kong, will be described in more detail below.

In the UK in September 2017, the UK Department of Business, Energy and Industrial Strategy (BEIS) launched the Green Finance Initiative (GFI) Taskforce to encourage the growth of the green finance sector in the UK's transition to a low carbon economy. The GFI report has supported TCFD implementation in UK as part of accelerating the low carbon transition process in UK. The key reason is that in order to accelerate new green investments, via debt or equity, the leading companies, banks and markets will all need to be able to understand the relevant climate risks and implications. Hence clear robust and transparent climate risk reporting will be key. This has resulted in the UK Government voicing its support for the TCFD recommendations and promoting the British Standards Institution's (BSI) work on an International Voluntary Sustainable Financial Management Standard, which is due to come out soon.

Green finance and climate governance have also been growing steadily in China. The concepts of ESG (environmental, social and corporate governance), TCFD and impact investment, have also being increasingly accepted and supported by the PRC Government, regulators, financial institutions and listed companies. In particular there has been strong support for ESG and TCFD adaptation for China SOEs and banks. There has also been good international cooperation between China and the City of London in UK. A good example is the successful UK China TCFD Pilot Program established between leading Chinese and UK companies.

In May 2018, the Hong Kong Stock Exchange published its first report on the analysis of environmental, social and governance (ESG) practice disclosures which are closely linked to TCFD recommendations. They have reviewed 400 annual reports of listed issuers and reported that that only about one-third, some 38 per cent, of the listed issuers and companies under review have been in full compliance with the 11 aspects of the environmental, social and governance areas which are subjected to ESG disclosures. There is much room for significant improvements by these listed companies.

The Hong Kong Stock Exchange has urged their listed companies and issuers to improve and enhance their disclosures in their new ESG reports.

Earlier in 2018, the European Commission had also published new legislative proposals on sustainable green finance. These new measures have major impacts on asset managers, banks and companies. They have to integrate ESG factors and risks in their corporate management. There will be new formal obligations to disclose how sustainability risks are integrated into organisation management and services provided to clients. Firms should pursue low carbon emission objectives and comply with new 'low carbon' or 'positive carbon impact' benchmarks.

Looking ahead, with the looming Paris Agreement commitments and plans by key governments globally to improve climate governance, there will be a strong possibility that 'voluntary reporting' could become mandatory in the near future. It is recommended that the boards and senior management of leading banks and companies globally should develop good understanding of how they can improve their climate-related financial disclosures accordingly to meet the new requirements. This would be necessary in order for them to to shield their corporations from potential challenges and lawsuits. In addition, these should also help them to better seize and monetise the new climate-related opportunities that might be available to their businesses plus should help leading banks and companies to avoid rising challenges and even lawsuits from their regulators, lenders, shareholders and ethical investors.

The details of the UK GFI governance improvements and the Hong Kong Stock Exchange ESG improvements will be discussed in two case studies below. Details of recent ethical investors' serious challenges to leading companies will also be discussed.

UK GFI governance improvements case study

In the UK, financial stability risks have been a key concern for the Bank of England. It has led the G20 TCFD Study and has recommended TCFD implementation to improve climate risk reporting in the UK. This will improve government and regulators meeting monetary and financial stability objectives. Delayed implementation of the reporting requirements on material emerging issues would allow financial risks to build up to unmanageable levels. The speeds at which financial instability and market re-pricing could occur are uncertain. However these could seriously affect the UK's financial stability and reputation (Bank of England, 2017).

The trust levels that UK investors and stakeholders should have in the UK capital markets are considered paramount. The expectations for UK capital markets to handle the rising climate and sustainability challenges are increasing. Improving disclosure and transparency would help to improve trust in the capital markets. In addition, companies and their directors might face increasing legal liability exposures by failing to assess and manage environmental risks in accordance with their duties or failing to report risks. Without credible comparability of climate related disclosures, based on a level playing field of disclosure requirements, it would be difficult for UK financial system regulators, investors and capital markets to review the information reported by the banks and companies. UK business productivity and performance could face serious risks, without accurate information. The implementation of climate and sustainability practices at the organisation level could therefore help companies to become more competitive and robust. Hence there is strong support in the UK Green Finance team for adaptation of the TCFD recommendations in UK (GFI, 2017).

The UK Government has also been supporting climate finance and governance growths. BEIS and HM Treasury have jointly launched a new UK Government plan in 2017 to accelerate the growth of climate finance in the UK. They want to continue building on UK strengths as one of the world's largest financial centres and an attractive place to do business, despite Brexit. There are good professional depths of innovation, products and capital on new green finance and investment. In addition, there is good legal expertise and emerging markets knowledge in the City of London plus other financial centres in the UK. Hence it is very important to ensure that UK companies provide the best and most reliable information on climate-related risks and strategies. These would be fundamental to UK continuing to act as one of the global green financial centres (City of London Corporation, 2017).

It is generally believed that leadership in the provision of climate and sustainability information for financial decision-making, enabled by effective disclosure and reporting frameworks, would be a critical competitive advantage for the UK. Disclosure and transparency would also help to improve trust in capital markets. These will be important for London's international reputation and its attractiveness to issuers on the London Stock Exchange. The Bank of England has led the discussion globally on financial stability risks flowing from climate change and has a position of authority on the development of the sustainability aspects of the financial stability regime globally. The UK has a good base of regulations and guidance on sustainability issues which could be further improved to properly include climate-related disclosure obligations (TheCityUK, 2017).

The global competitive financial landscapes have been changing rapidly. A growing number of global financial centres supported by their proactive policymakers, have started to challenge London. Good examples include Paris, Dublin and Frankfurt, which have all been actively competing with London to become future financial centres of Europe, especially after Brexit (UK Parliament, 2018).

Despite wide acceptance in the UK business and finance sectors of the need to reduce emissions, there have been information shortcomings and failures on the financial risks and opportunities which have hampered decision making. A good example is that assumptions about the severity and implications of climate risks have varied widely between corporates.

There have also been limited forward-looking climate-related financial information disclosures by companies. ClientEarth has also reported there have been some serious disclosure failures in the London Stock Exchange. They have reported several UK listed companies to the Financial Reporting Council as these UK companies have not met the disclosure requirements set out under the 2006 UK Companies Act (ClientEarth, 2016).

UK companies and banks should use the TCFD framework to develop their financial, corporate governance and stewardship disclosures on a comply or explain basis. There will be a comprehensive effort by the UK Government and relevant UK Financial regulators to support successful adoption, implementation, and enforcement of the TCFD recommendations. These would include public disclosure, public rankings, off-the-shelf tools and scenarios plus publicly available datasets. The UK Government is considering conducting a general review of all climate risk financial disclosures by UK companies and banks in 2020. This should help to monitor and encourage market adoption amongst both issuers and users. If adoption is deemed insufficient, the Government would consider to what extent further measures would be required to encourage take-up.

Relevant UK financial regulators will also be integrating the relevant TCFD recommendations throughout the existing UK corporate governance and stewardship reporting framework. It is expected that the UK Government and financial regulators will be creating and publishing guidelines and referencing these in the corpus of relevant UK regulations.

The new UK guidelines would clarify certain TCFD recommendations to make them more readily implementable. A good example is better definitions for physical climate scenario analysis and the disclosure of assumptions. Key to this process would be for relevant regulators to identify the most effective mode by which the guidelines could be accommodated within its regulatory remit. In the event that a regulator should determine that the guidelines cannot be accommodated within its remit, then the mandate should be amended accordingly.

To ensure that the TCFD recommendations would be properly integrated into the UK corporate governance and reporting framework, the guidelines will need to be appropriately referenced in the corpus of relevant financial rules, codes and guidance such as the UK FCA Handbook plus corporate governance codes and stewardship codes. The guidelines should include each of the four key areas of the TCFD recommendations covering governance, strategy, risk management plus metrics and targets.

The new UK guidelines should define the disclosure requirements plus ensure that information would be disclosed on a consistent and transparent basis. The guidelines should make clear that the assumptions used for strategy planning, scenario formulations or projections should also be disclosed by companies. It should ensure that companies provide scenario-based disclosures of how their business strategies and financial planning might be affected by climate-related risks. It should also ensure that companies report new revenues from green business areas plus take account of how different jurisdictions would respond to new disclosure needs (GFI, 2017).

The new UK guidelines should also define which companies would be covered by the new disclosure requirements. This would be in large part determined by the scope of existing rules and codes. Financial regulators will need to assess the suitability of the existing rules, codes and guidance. If these are not suitable, then new guidance should be considered.

In addition to the need for companies to comply with the new climate-related disclosures, there are several additional steps that the private and corporate sectors could also take. These would include providing appropriate training for staff to ensure sufficient organisational competence to evaluate climate risks plus formulate climate strategy, scenarios and targets.

UK government and financial regulators must clarify in their new guidelines that disclosing material environmental risks, including physical and transition climate-related risks, is already mandatory under existing law and practice. The guidelines should also clarify that the TCFD recommendations, given that they reflect key international norms for disclosing material climate-related risks, would enable companies to fulfil their legal obligations and duties. The TCFD recommendations for disclosing climate-related risk should be embedded in all relevant UK corporate governance and reporting frameworks. Appropriate revisions to UK legislation should be considered to further integrate the TCFD recommendations into the corporate governance and reporting framework. This will provide an opportunity to harmonise requirements across different sectors and make provisions for further accountability.

By implementing the various measures above, the UK should become the first jurisdiction globally to fully implement the TCFD. This would enable investors in the UK to differentiate between companies and assets that are aligned with the Paris Agreement and the UK's Clean Growth Strategy. These should have the effect of ensuring that companies which have implemented appropriate climate strategies and sustainability practices would have a lower cost of capital, higher productivity and better stock market performances. There should also be additional macro-economic benefits, including improve the UK economy productivity and putting UK industries at the leading edge of a global low carbon economic transformation (Clark, Feiner & Viehs, 2016).

Stock exchange case study: Hong Kong Stock Exchange ESG requirements

Leading international stock exchanges have put high priority on financial risks and stability management. Financial stability risks have been one of the top key concerns for all financial markets and stock exchanges globally. Climate financial risks plus climate change physical and transition risks are generally considered to be some of the key financial stability risks globally. Better climate risk disclosure and transparency are considered key to encouraging improved performance and trust in stock markets globally. Hence international stock markets have been including new reporting, such as ESG and/or TCFD, for their listed companies to improve reporting on their climate risks, strategy and governance systems.

An international stock exchange example is the push by the Hong Kong Stock Exchange (HKEX) to improve the reporting of ESG disclosures by all the issuers and listed companies on the HKEX. On 18 May 2018, the HKEX published its first report on the analysis of ESG practice disclosure. The HKEX has reviewed 400 annual reports of their listed companies. HKEX reported that over 61 per cent of the listed issuers under review had the ESG reports incorporated in their annual reports and 39 per cent of them had standalone ESG reports. However only about one third, some 38 per cent, of the listed companies under review were in full compliance with the 11 aspects of the environmental and social areas subject to ESG disclosures. They have identified a few important areas for further improvements. HKEX has asked their listed issuers to consider enhancing their disclosures in their next ESG reports (HKEX, 2018).

HKEX has recommended all the listed corporates should establish a special ESG working group comprising senior management and other staff members who have sufficient ESG knowledge to conduct materiality assessments. The ESG working group should have clear terms of reference and should report to the Board.

HKEX has suggested all listed corporates should state, at the start of their ESG report, their commitment to ESG. They should also explain their management approach plus explain how these relate to the issuer's business. In addition, the ESG report should include the board's evaluation and determination of ESG risks plus how they would ensure that effective ESG risk management and internal control systems have been put in place.

HKEX has stressed that one of the key reporting principles would be materiality. All listed companies must consider the ESG aspects to be material to their business. They stressed that the process of stakeholder engagement is central to the assessment of materiality. Issuers would have to disclose in their ESG reports and explain how they have arrived at the conclusion that a particular aspect is or is not considered to be material.

Another key reporting principle is that on quantitative consistency. All companies should use measurable key performance indicators (KPIs) with consistent methodologies to allow for meaningful comparisons of ESG data over time. HKEX has suggested that all KPIs will need to be accompanied by a narrative statement which explains its purpose and impacts plus comparative data where available.

HKEX has stressed that the failure to report the reasons for non-disclosure of a particular ESG aspect would constitute a breach of the Listing Rules. Companies would need to check if they have any such omissions in their next ESG reports. Vague or partial disclosures are considered 'non-compliant'.

Companies should provide sufficient information about their corporate policies so investors would have better understanding. In addition, if any ESG aspect calls for information on several areas, then suitable disclosures should be required to be made in the ESG report.

All listed companies should ensure that they are in compliance with all relevant laws and regulations. Their ESG report should specify the relevant laws and regulations, as well as ways in which the issuer has ensured compliance. In the unlikely event that there are no relevant laws and regulations that have significant impacts, then clear reasons must be given by the company in their report to avoid breach of the Listing Rule.

Global investor challenges and implications for multinationals

There are rising concerns from ethical investments and shareholders globally about the implications of climate change on leading banks and companies globally plus how prepared they would be able to handle these. These have resulted in rising shareholder and investor challenges on leading companies and banks to improve their climate financial risk reporting and transparency.

In the face of all these intense international, government and shareholder pressures, leading banks and companies globally have to take these new requirements on climate-related financial disclosures very seriously. Their board and top management have to put in place the necessary new systems and processes to compile with the new disclosure and reporting requirements. Failure to do so could open the company and themselves to serious challenges and even lawsuits from the regulators, shareholders, investors and media.

A good shareholder action example is the important shareholders challenges to the ExxonMobil Board recently. Institutional investor coalitions and shareholder advocacy groups have been seriously challenging the ExxonMobil's board on the company's recent weak climate risk report and demanding improvements. The shareholders have been led in their challenges by the office of New York State Comptroller. They considered the company's recent climate report to be defective and unresponsive to a resolution from the shareholders (IEEFA, 2018).

In the face of rising global climate change risks and impacts, ExxonMobil shareholders have filed many resolutions with the company requesting disclosures on the impact of these changes on the company's reserves and resources plus on the associated financial risks. Last year, 62 percent of ExxonMobil shareholders approved a resolution which pressed specifically for ExxonMobil to prepare a climate report assessing the risks from a '2-degree scenario', which stands for the scenario for a global warming of 2 degrees Celsius over preindustrial levels.

These disputes between ExxonMobil and its investors have led the US Securities and Exchange Commission to side with shareholder requests to be allowed to vote to request a climate risk report in March 2016. Since then

Exxon's stock has dropped 11 per cent while the US Standard & Poor's 500 Index has risen 32 per cent. The stock market and investors seemed to have delivered their summary judgment on ExxonMobil.

ExxonMobil's climate report published recently has said that potential climate-change impacts would be unlikely. They have concluded that even if demands fall or regulatory restrictions would be enacted, then neither would pose serious material risks to its reserve calculations or to its financial-risk profile. Their report has assumed a business-as-usual operational scenario. The report did acknowledge the various climate risks globally. However no responsibility was assigned to the company in creating these risks, despite the high GHG emissions generated from their oil and gas operations globally for many decades. In addition, they also do not see any significant roles for ExxonMobil in mitigating these serious climate risks. ExxonMobil has been planning to continue increasing capital expenditures to expand their oil and gas operations in the USA and abroad.

The shareholders felt they have the right to be critical of the substantive content of the report as they considered that there were many shortcomings and flaws. Many of the shareholders, including the New York Common Fund, saw the report as overly general and unrealistically reliant on optimistic assumptions. In addition, ExxonMobil continued denial of the risks that climate change will pose to its business is contrary to the positions of other leading international oil and gas companies.

The shareholders have challenged ExxonMobil, stating that there were many flaws in the report, including the failure to place the climate issue into a financial risk context. It has failed to include the current level of baseline financial risks facing the company or layer in carbon restrictions, as other company's climate risk reports have done. The report has failed to give a complete quantitative picture of the company's reserves, proven or unproven under a 2 degree Celsius scenario. It has also omitted any extended treatment or discussions of reserve valuations. It also did not address any potential global climate change policies and risks. It avoided mentioning investigations by the SEC and two state attorney generals into how the company has been accounting for its assets. The report has also failed in addressing some of the material financial losses that the company has suffered in its oil and gas business over the past decade. These included the 2016 write-off of more than four billion barrels of reserves in the Canadian tar sands and overpayment for the reserves secured in the $6 billion acquisition of XTO natural gas assets.

The ExxonMobil shareholders and investors have said they wanted clear empirical understandings of the company's production assumptions, annual-use assumptions, current and projected acquisitions and other additions to its proven reserve category, plus various other relevant economic and financial projections. The absence of this vital information had placed ExxonMobil's physical and financial reserve calculations in serious question. The report has also provided no basis for the company's conclusion that its reserves and resources would be unaffected by potential climate and carbon constraints.

In response to the shareholder actions, ExxonMobil has, in late January 2019, written to the US Securities and Exchange Commission (SEC), to say that the management considered the investor proposals to be misleading and was an attempt to micro-manage the company. Both sides have been awaiting US SEC ruling prior to the important vote at ExxonMobil's May 2019 annual meeting (Reuters, 2019).

Looking ahead, with the current global focuses on climate change and global warming, it is important for business leaders to understand that companies cannot expect their shareholders just to accept, without questions, company statements and reports without good supporting evidence. This is especially the case when a company's financial reporting is under scrutiny and the company's recent financial performance is raising so many red flags. The current investment climate will require that the management of Exxon-Mobil and other leading companies carefully weigh their tradition claims to proprietary secrecy against investor demands for greater transparency and disclosures, especially in relationship to climate change.

All the boards globally have to understand that when a company remains unresponsive to investors, then the exercise of strong shareholder rights is in order. Shareholders have two immediate paths they can take. They can simply cast a 'no' vote for one or all of the company's proposed board of directors to reflect shareholders' dissatisfaction with the report and the underlying views it reflects. Alternatively they can invoke proxy access rights that will allow certain large shareholders to offer candidates for board consideration. Both of these shareholder actions would result in serious problems for the board and significant damages to the company's reputation globally.

Leading international investment banks and asset managers globally have also been enhancing their sustainability investment strategy on the huge investment portfolios that they are managing globally. Many leading investment banks and fund managers have been setting out in more clear terms how they plan to engage with and challenge the companies that they are investing in, on their climate change strategies. Many have said that they will be integrating ESG factors into their investment evaluations, decision processes and reporting.

A good example is that multinational investor BNP Paribas Asset Management (BNPP AM) has improved its sustainable investment strategy recently. It has set out in detailed how it intends to align its entire €399 billion (U$ 447 billion) portfolio with the goals of the Paris Agreement by 2025. BNPP AM has just issued its new Global Sustainability Strategy in March 2019. The strategy sets out how BNPP AM plans to engage with companies in its portfolio, including fossil fuel firms, on climate issues and governance. Under the new strategy, every company that the BNPP asset manager invests in will need to demonstrate how it is working towards the Paris Agreement's 2 degree Celsius goal by 2025. BNPP AM has said that they plan to integrate ESG factors into all of their investment processes, including investment research, risk management, voting and reporting. They have set up an ESG validation committee to oversee the process across the firm. They have also expanded its ESG expert team plus set up new ESG focused training programs for all their staff (BNPP AM, Sustainability Roadmap to Deliver Paris-aligned Investment Portfolio by 2025, 2019).

BNPP AM has also recently revised their enhanced coal investment policy. They have stated that from 2020 it will exclude from its investment portfolios, companies which have been deriving more than 10 per cent of their revenue from mining thermal coal. They will also not invest in major coal companies which accounted for one per cent or more of total coal global productions. Coal-fired power generators with a carbon intensity above a certain threshold will also be excluded. BNPP AM has also planned to increase its engagement with companies in other carbon intensive industries, such as oil and gas, steel and cement (BNPP AM, Revised Coal Investment Policy, 2019).

Another major global employee-owned asset manager, Neuberger Berman, has also launched similar new climate investment strategies. They have announced in March 2019 that they will be launching a new climate strategy which will aim to better analyse the climate risks facing its U\$300 billion global investment portfolio.

These international cases demonstrate the high importance that ethical investors, asset managers and stakeholders are placing on the climate change risks and their impacts on companies globally. Management of leading companies have to take these concerns into serious consideration in planning their business investments and sustainable future growths globally. Otherwise they could be facing serious challenges and lawsuits from their investors, regulators, stakeholders and media.

Bibliography

ABC News, South Australia Biogas Human Waste Pilot Plant, USA, April 2019.

ACS (American Chemical Society), Selective Electrochemical Reduction of Carbon Dioxide Using Cu Based Metal Organic Framework for CO2 Capture, USA, 2018.

Africa Development Forum, Electricity Access in Sub Saharan Africa, World Bank, USA, 2019.

Agriculture and Agri-Food Canada, Holos Climate Change Agriculture Model, Government of Canada, Canada, 2018.

Agriculture and Agri-Food Canada, Climate Change and Agriculture, Government of Canada, Canada, 2018.

Albright, R., et al., Reversal of Ocean Acidification Enhances Net Coral Reef Calcification, *Nature*, 531, pp. 362–365, USA, 17 March 2016.

Alini, E., Worried about the Climate and Carbon Taxes? What to Know about Ottawa's New Electric-vehicle Incentive, *Global News*, Canada, April 2019.

Amadeo, K., Carbon Tax, Its Purpose, and How It Works, USA, June 2019.

APEC, Life Cycle Assessment of Photovoltaic Systems in the APEC Region Report, Singapore, April 2019.

Asian Development Bank, ADB Signs Landmark Project with Icelandic, Chinese Venture to Promote Zero-Emissions Heating, Manila, 2018.

Asian Development Bank, Green Cities, Manila, 2012.

Assessment Agency (EU EDGAR), Emission Database for Global Atmospheric Research, Belgium, 2011.

Bank of England, Quarterly Bulletin Q2 Article: The Bank of England's Response to Climate Change, UK, 2017.

Barton, D., Half a Billion: China's Middle-class Consumers, Global MD of McKinsey & Co, *The Diplomat*, USA, 30 May 2013.

BBC, How Will Our Future Cities Look? BBC London UK, 17 February 2013.

BBC Environment, Carbon 'Bubble' Could Cost Global Economy Trillions, UK, June 2018.

BBC News, Climate Change Impacts Women More than Men, UK, 8 March 2018.

BBC News, Stephen Hawking's Warnings: What He Predicted for the Future, Paul Rincon, Science Editor, BBC UK, 15 March 2018.

Bell, V., Air Pollution is a Bigger Killer than Smoking, *Daily Mail*, 12 March 2019.

Berger, J., Copenhagen Striving To Be Carbon Neutral: The Economic Payoffs?, UK, July 2017.

Bloomberg, China Boosts Solar Target for 2015 as It Fights Pollution, USA, 2015.

Bloomberg, Beijing to Shut all Major Power Plants to Cut Pollution, USA, 2015.

Bloomberg, All Forecasts Signal Accelerating Demand for Electric Cars, USA, July 2017.

Bloomberg, India's Blue Sky Pledge Gives Power to Country's Green Bonds, Anindya Upadhyay, USA, 24 July 2017.

Bloomberg, Tesla Finishes First Solar Roofs, USA, August 2017.

BNEF (Bloomberg New Energy Finance), The Future of China's Power Sector. From Centralised and Coal Powered to Distributed and Renewable? USA, 14 October 2013.

BNEF (Bloomberg New Energy Finance), Electric Vehicle EV Battery Pack Cost Forecasts, USA, 2018.

BNEF (Bloomberg New Energy Finance), Bullard: Tech Investments Are Powering Up Clean Energy, USA, October 2018.

BNEF (Bloomberg New Energy Finance), Battery Power's Latest Plunge in Costs Threatens Coal, Gas, USA 26 March 2019.

BNPP AM (BNP Paribas Asset Management), Revised Coal Investment Policy, Paris, 2019.

BNPP AM (BNP Paribas Asset Management), Sustainability Roadmap to Deliver Paris-aligned Investment Portfolio by 2025, Paris, 2019.

Boao Forum for Asia, 2014 Forum Speeches & Proceedings, BFA, Hainan, PRC, 2014.

Boao Forum for Asia, 2015 Forum Speeches & Proceedings, BFA Hainan, PRC, 2015.

BP, Carbon Neutral Management Plans, London, 2017.

C40 City & Arup, Deadline-2020 City Report, UK, 2018.

Canada Ecofiscal Commission Report on Carbon Pricing, Canada, April 2018.

Carbon Brief, Food & Farming, Rise in Insect Pests under Climate Change to Hit Crop Yields, Study Says, August 2018.

Carbon Tracker Initiative UK, The Carbon Majors Report, London UK, 2017.

CDP (Carbon Disclosure Project), Global City Report, UK, 2016.

CDP (Carbon Disclosure Project) & CAI (Climate Accountability Institute), The Annual Carbon Majors Database Report for 2018, UK, 2018.

CGIAR, Fighting Floods by Sponge Cities, Udon Thani Thailand, Thailand, 2018.

Chartered Management Institute, 'Winning Ideas – Top 5 Management Articles of the Year', London UK, February 2016.

China Carbon Forum, China Carbon Pricing Survey, Beijing, PRC, 2017.

China Dialogue, Five Things to Know about the China National Carbon Market, China, December 2017.

China Dialogue, Expert Roundtable: Is China Still on Track to Reach Its Paris Targets?, China, June 2018.

China National Bureau of Statistics, Statistical Communiqué of the People's Republic of China on the 2014 National Economic and Social Development, China, 2015.

Christian Aid, Counting the Cost: A Year of Climate Breakdown, UK, December 2018.

CIFOR (Center for International Forestry Research), Forests and Climate Change, Indonesia, 2019.

City of London Corporation, Total Tax Contribution of UK Financial Services, UK, 2017.

Clark, G., Feiner, A. & Viehs, M., From the Stockholder to the Stakeholder: How Sustainability Can Drive Financial Outperformance, Smith School of Enterprise and the Environment, Oxford University, UK, March 2015.

CleanTechnica, World's First Advanced Offshore Wind Power Radar System Now Operational, USA, 2016.

Climate Policy Info Hub, The Global Rise of Emissions Trading, EU, 2018.

ClientEarth, Review of UK Companies' Climate Disclosures, UK, 2016.

Climate Central, John Upton, China, India Become Climate Leaders as West Falters, Canada. April 2017.

Climate Action Tracker, China, US and EU Post-2020 Plans Reduce Projected Warming, USA, 2013.

CNRS (Centre national de la recherche scientifique), Climate-driven Range Shifts of the King Penguin in a Fragmented Ecosystem, Paris, France, 2017.

Coglianese, C. & Nevitt, M., Actually, the United States Already Has a Carbon Tax, *Washington Post*, January, 2019.

Columbia University, Committee on Global Thought, USA, 2019.

Comiso, J.C & Hall, D.K. Climate Trends in Arctic as Observed from Space, WIREs Climate Change, USA 2014.

Computer Weekly, Google to hit 100% Renewable Energy Target for Datacentres in 2017, USA, 2016.

Computer Weekly, Telefónica Increases Use of Renewable Energy to Fight Climate Change, USA, 2017.

CSO Energy Acuity, Global Renewables Top Ten Companies, UK, 2019.

Cuéllar-Franca, R. & Azapagic, A., Journal of CO2 Utilisation, Carbon Capture Storage & Utilisation Technologies: A Critical Analysis and Comparison of Their Life Cycle Environmental Impacts, USA, March 2015.

DuPont, Sustainable Energy for a Growing China, May 2013, USA, 2013.

Dutch Marine Energy Centre, FORESEA Funding Ocean Renewable Energy through Strategic European Action Project, Netherlands, 2019.

Economist Intelligence Unit, The Cost of Inaction: Recognising the Value at Risk from Climate Change, UK, 2015.

EEA (European Environmental Agency), Agriculture and Climate Change, Brussels, 2018.

Efthymiopoulos, I., Recovery of Lipids from Spent Coffee Grounds for Use as a Biofuel, PhD thesis, UCL, London, UK, 2018.

EIA, Explosion of HFC-23 super greenhouse gases is expected, USA, 2015.

Elsner, Kossin & Jagger, T.H. The Increasing Intensity of the Strongest Tropical Cyclones, *Nature*, 455, no. 7209, USA, 2008.

Energy Narrative, Is 2¢ a kWh Solar Power Real? Jed Bailey, USA, 15 February 2018.

Energy Foundation, Statement on China's Launch of the National Emissions Trading System, USA, December 2017.

Energy Research Institute, China 2050 High Renewable Energy Penetration Scenario and Roadmap Study, USA, 2015.

Environmental Defence Fund, Five Reasons to Be Optimistic about China's New Carbon Market, USA December 2017.

EU ESCO Committee of China Energy Conversation Association. 'Notice on the Disposal of Hydrofluorocarbon', European Commission, Joint Research Centre (JRC)/Netherlands Environmental, Brussels, July 2015.

EU, Framework for State Aid for Research and Development and Innovation, Brussels, 2014.

EU, An EU Strategy on Heating and Cooling, Brussels, 2016.

EU, New Renewable Energy Directive to 2030, Brussels, November, 2016.

EU Environment Agency EEA, Renewable Energy in Europe Report, Brussels, 2017.

EU, Clean Energy for all Europeans, Brussels, 2018.

European Commission, CCS Directive, Brussels, 2012.

European Commission, Causes of Climate Change, Brussels, 2019.

Fenby, Jonathan, *Will China Dominate the 21st Century?*, London, UK, 2014.

Financial Post Canada, Ottawa to Return 90% of Money It Collects from Carbon Tax to the Canadians Who Pay It, Canada, 2018.

Fine, M. and Tchernov, D., Scleractinian Coral Species Survive and Recover from Decalcification, *Science*, USA, 2006.

Fisher, R., Ury, W., Patton B., *Getting to Yes: Negotiating Agreement Without Giving In.* New York, NY, Penguin Books, USA, 1991.

G20 TCFD, Task Force on Climate-related Financial Disclosures (TCFD) Report, USA, July 2017.

Gabbatiss, J., CO2 levels expected to rise rapidly in 2019, UK Met Office scientists warn, *Independent*, 25 January.

GFI (Green Finance Initiative), Green Finance Initiative Report, London, UK, 2017.

Glick, Daniel, The Big Thaw, *National Geographic*, USA, 2019.

Global Commission on the Economy and Climate, The New Climate Economy Report: Better Growth Better Climate, USA, 2014.

Global Commission on the Economy and Climate, The New Climate Economy Annual Report, USA, 2017.

Global Commission on the Economy and Climate, The New Climate Economy Report: The 2018 Report of the Global Commission on the Economy and Climate, USA, 2018.

Global CCS Institute, Global CCS Report, London UK, 2015.

Global CCS Institute, Yanchang Petroleum Report 1: Capturing CO2 from Coal to Chemical Process, UK, 2015.

GPCA and McKinsey Report 'Thoughts for a New Age in Middle East Petrochemicals' released at 10th GPCA Forum, Dubai, UAE, November 2015.

Green, F. & Stern, N., China's 'New Normal': Structural Change, Better Growth and Peak Emissions, LSE, UK, March 2015.

Greenwich Council, New E-Car Club for Low Emission Neighbourhood, UK, 2019.

Guardian, London Energy Customers Locked into District Heating Systems, UK, 2017.

Guo Peiyuan, SynTao, The Concept of Responsible Investment Is Increasingly Recognized and Accepted by Chinese Financial Institutions, Beijing, PRC, March 2019.

Haerens, J., The Breakthrough of CCS/CCU: An Analysis of Drivers and Hurdles, Master's thesis, University of Ghent, The Netherlands, 2017.

Harvard Business School, Should You Make the First Offer?, USA, July 2004.

HKEX (Hong Kong Stock Exchange), Report on the Analysis of Environmental, Social and Governance (ESG) Practice Disclosure, Hong Kong, 18 May 2018.

Howell, Lord David, Oil and Money, a Combo that Faces a Cloudy Future, *Japan Times*, Japan, October 2018.

Huidian Research, Indepth Research and Forecast of China Ethylene Industry for 2013–2017, Beijing, PRC, 2013.

Hunter, Estimating Sea level extremes under conditions of uncertain sea level rises, *Climate Change* 99, nos 3–4, USA, 2010.

ICAP (International Carbon Action Partnership), Lessons Learnt from Chinese Pilot ETS Cap Setting and Allowance Allocation, UK, 2014.

ICIS, Global Annual Base Oil Conference, London, UK, February 2018.

Idso, Craig, CO2, Global Warming & Coral Reefs: Prospects for the Future, USA, 2009.

IEA, Energy Technology Perspectives, International Energy Agency, Paris, 2010.

IEA, World Oil Market Report 2012, International Energy Agency, Paris 2012.

IEA, Report & Roadmap for Energy Conservation & GHG Emission Reductions by Catalytic Processes, IEA, Paris, 2013.

IEA, Energy Balances. International Energy Agency, Paris, 2014.

IEA, Energy Technology Perspectives, International Energy Agency, Paris, 2014.

IEA, World Energy Outlook 2014, International Energy Agency, Paris, 2014.

IEA, World Energy Outlook 2015. International Energy Agency, Paris, 2015.

IEA, World CO2 Emissions from Fuel Combustion: Database Documentation, International Energy Agency, Paris, 2015.

IEA, World Energy Outlook, IEA WEO Report, Paris, 2017.

IEA, WEO Special Air Pollution Report, Paris, 2017.

IEA, Global Energy and CO2 Status Report of 2018, Paris, March 2019.

IEEFA, IEEFA Update: ExxonMobil Empty Climate Risk Report: A Majority of Shareholders Want – and Deserve – More Transparency, USA, April 2018.

IMF (International Monetary Fund), Fiscal Implications of Climate Change, Fiscal Affairs Department, USA, March 2008.

IMF (International Monetary Fund), World Economic Outlook Database. International Monetary Fund, Washington D.C., USA, 2015.

Imperial College Digital Economy Lab, Digital City Exchange, London, UK, 2019.

Imperial College London & ICROA, Unlocking the Hidden Value of Carbon Offsetting, UK, 2014.

Imperial College London, Climate Risks and Investment Impacts, Charles Donovan, Director at the Centre for Climate Finance and Investment, Imperial College London Business School, UK, 2018.

Imperial College London, BECCS Deployment Report, London, UK, 2018.

Imperial College London, Climate Change and Environmental Pollution Affects Heart and Lung Health, Explain Experts, Maxine Myers, Joy Tennant, Martin Sayers, UK, 12 March 2018.

Imperial College London, BECCS Deployment: A Reality Check, Fajardy et al., Grantham Institute Briefing paper No. 28, UK, January 2019.

INSEAD, AI, Business, Organizations, Society: Holistic Approach to the Humans+ Machines Loop, Theos Evgeniou, France, 2019.

IOT Agenda, Using Technology to Save Nature, USA, 2017.

IRENA, Biofuels for Aviation, Abu Dhabi, 2017.

IRENA, Investments in Renewables Analysis, Abu Dhabi, 2018.

Japanese Government, Blueprint towards a Possible Hydrogen Society by 2040, Tokyo, 2017.

King Abdullah Petroleum Studies & Research Centre KAPSARC 2014 Discussion Paper on Lowering Saudi Arabia's Fuel Consumption & Energy System Costs without Increasing End Consumer Prices, KAPSARC Riyadh, Saudi Arabia, March 2014.

KPMG, Investment in PR China Report, UK, April 2012.

LanzaTech, Biological Conversion of Carbon to Products through Gas Fermentation, New Zealand, 2019.

LSE Grantham Institute, Post-COP24 Forum, London, December 2018.

Marine Energy.biz, Third MaRINET2 Call Provides €1.2M Testing Boost for Renewables, UK, 2019.

McKinsey, How Companies Can Adapt to Climate Change, USA, July 2015.

McKinsey, Disruptive Trends that Will Transform the Auto Industry, USA, January 2016.

McKinsey, How Solar Energy can Finally Create Values, USA, October 2016.

McKinsey, Competing in a World of Sectors Without Borders, USA, 2017.

McKinsey, Smart City Developments and Improvements. USA, March 2018.

McKinsey, Bringing Solar Power to the People, USA, June 2018.

McKinsey Consulting, Bringing Solar Power to the People, USA, June 2018.

McKinsey, The Future of Electricity Rate Design, USA, March 2019.

McKinsey, Global Energy Perspective, USA, August 2019.

McKinsey Quarterly, The Disruptive Power of Solar Power, David Frankel, Kenneth Ostrowski, and Dickon Pinner, USA, April 2014.

McKinsey Quarterly, Peering into Energy's Crystal Ball, Scott Nyquist, McKinsey, USA, July 2015.

McKinsey Quarterly, Battery Technology Charges Ahead, Russell Hensley, John Newman, and Matt Rogers, USA, March 2018.

Mining.com, Scandinavian Biopower to Invest in a Biocoal Plant in Mikkeli Finland, UK, 2016.

MIT, Climate Forecasts, USA, 2017.

MIT, MIT Climate Action, USA, 2017.

MIT, Senseable Cities Lab, USA, 2019.

Nagoya University, Evaluating the Contribution of Black Carbon to Climate Change, Japan, 2018.

NASA, What Is Climate Change?, USA, May 2014.

NASA, Global Climate Change Sea Level Rise Data, USA, 2019.

NASA Jet Propulsion Lab & CIT, Key Indicator Global Climate Change, USA, August 2014.

National Research Council, Ecological Impacts of Climate Change, Washington USA 2008.

National Resources Defense Council, The Consequences of Global Warming on Glaciers and Sea Levels, USA, August 2014.

NCA (National Climate Assessment), USA Fourth National Climate Assessment, USA, 2018.

Nerem, Proceedings of the National Academy of Sciences on Sea level Rises, USA, 2018.

New York Times, How to Play Well with China on USA President Obama & PRC President Xi Jingping Meetings in June 2013, *New York Times*, New York, USA, 2013.

NOAA, Climate Change: Atmospheric Carbon Dioxide, USA, 2018.

NRDC, Global Warming, USA, March 2016.

NSIDC (National Snow & Ice Data Centre), 2018 Winter Arctic Ice Report, USA, 2018.

NSIDC (National Snow & Ice Data Centre), Glaciers and Climate Change, USA, 2019.

OECD, Green Growth Declaration, Paris, France, 2009.

OECD, The Greening of Agriculture, Agricultural Innovation and Sustainable Growth, Paper prepared for the OECD synthesis report on agriculture and green growth, Paris, November 2010.

OECD MENA, Task Force on Energy & Infrastructure 2013 Report on Renewable Energies in the Middle East and North Africa MENA: Policies to Support Private Investment, with inputs by Henry Wang and other OECD MENA Task Force team members, OECD, Paris, 2013.

OPEC, World Oil Outlooks of 2013, Vienna, Austria, 2013.

OPEC, World Oil Outlooks of 2014, Vienna, Austria, 2014.

OPEC IEA IEF Energy Conference 23 March 2015, Riyadh, Saudi Arabia, 2015.

Parr, M., Diverting Fossil Fuel Investments to Renewables Is Not Enough, Euractiv, Brussels, April 2019.

PDO Petroleum Development Oman, Mirrah Solar Project, Oman, 2019.

Plasteurope.com, IKEA/ NESTE: Polyolefins from Bio-naphtha: Commercial-scale Pilot Plant to Start-up in Autumn, Belgium, 2018.

PNAS (Proceedings of the US National Academy of Sciences), Climate-change-driven Accelerated Sea-level Rises, USA, 2018.

Politico, 5 Takeways from COP24, Paris, USA, December 2018.

Power Engineering International, The Three Ds of Modern Power, USA, May 2017.

PRC, China's pledge to the Copenhagen Accord. Compilation of information on nationally appropriate mitigation actions to be implemented by Parties not included in Annex I to the Convention, Beijing, 2010.

PRC, China's 12th Five Year Plan, Twelfth Five-Year Guideline, 2011–2015, Beijing, 2011.

PRC, Second National Communication on Climate Change of The People's Republic of China, November, Beijing, 2012.

PRC, National Action Plan on Climate Change (2014–2020), Beijing, 2014.

PRC, Energy Development Strategy Action Plan (2014–2020), Beijing, 2014.

PRC, Technology Roadmap for Energy-saving Vehicles, Beijing, October 2016.

PRC, New FDI Measures: Management Measures for Approval and Filing of Foreign Direct Investment (FDI) 外商投资项目核准和备案管理办法, MOFCOM, Beijing, PRC, May 2014.

PRC Government, The Catalogue of Priority Industries for Foreign Investment in Central and Western China 中西部地区外商投资优势产业目录, PRC Government, Beijing, PRC, 2013.

PRC Government, 2013 Catalogue of Investment Projects Approved by Government (2013) 政府核准的投资项目目录 (2013年本), PRC Government, Beijing, PRC, 2013.

PricewaterhouseCoopers, China M&A 2012 Report, PWC UK, May 2013.

Princeton University, Earth's Oceans Have Absorbed 60 Percent More Heat than Previously Thought, USA, October 2018.

RE100, Report and Briefings, USA, 2019.

REN21, Renewable Energy Policy Networks for 21st Century, The Renewables 2017 Global Status Report, REN21 Secretariat, Paris, 2017.

REN21, Renewables 2017 Global Status Report, Paris, 2018.

Reuters, Exxon Asks U.S. Regulator to Block Climate-change Resolution: Investors, USA, 2019.

Ricke, K., Drouet, L., Caldeira, K., Tavoni, M. Country-level Social Cost of Carbon, *Nature Climate Change* 8, 895–900, September 2018.

Rik van dan Berge & Wang Henry, Report on Clean Coal Technology in China – A Strategy for Netherlands, Twente University, Netherlands, 2009.

Rogers, G., Planning a Successful TCFD Project, LinkedIn blog, USA, November 2018.

Royal Society, Climate Change Global Warming, London, UK, 2008.

Royal Society, Global Responses to Climate Change, London UK, 2014.

Sabine, C.L., The Oceanic Sink for Anthropogenic CO2, USA 2004.

SCMP, China's National Carbon Trading Rollout Expected to Have Major Impact on Key Industries, Hong Kong, April 2017.

SCOR, Report of the Ocean Acidification Group, Vienna, Austria, 2009.

Seatrade Maritime News, IMO 2020 Sulphur Regulation, UK, 2018.

Seymour, F. & Busch, J., *Why Forests? Why Now?* Center for Global Development, USA, 2016.

Shell International, Shell Sky Scenario, London, UK, 2018.

Slovenia Times, China's Path to a Green Economy, Slovenia, July 2017.

Smith, S., Environmentally Related Taxes and Tradable Permit Systems in Practice, OECD, Environment Directorate, Centre for Tax Policy and Administration, Paris, June 2008.

Stern, Lord Nicholas, Chair, Grantham Research Institute, London School of Economics and Political Science Speech, 'Post-COP24: Where Do We Go from Here?' at LSE Post-COP24 Forum, London, UK December 2018.

Sustainable Energy for All, Progress Toward Sustainable Energy 2015, UK, June 2015.

Techtarget, How Climate Change Threats Can Inform Cybersecurity Strategies, USA, 2018.

The National, China largest net importer of crude oil report 6 Mar, USA, 2013.

TheCityUK, Key Facts about the UK as an International Financial Centre, UK, 2017.

Toyota, Sustainability Environment Report 2018, Japan, 2018.

UK Department of Business, Energy and Industrial Strategy (BEIS), Smart City & Grid Developments Report, UK, July 2017.

UK Met Office, Climate Summaries 2018, London, UK, January 2019.

UK Met Office, UK Climate Projections 2018, UKCP18, London, UK, 2018.

UK Met Office, Warming: A Guide to Climate Change, UK, 2011.

UK Parliament, The Energy and Climate Change Committee, UK CCS Competition, House of Commons, UK, 2016.

UK Parliament, Electric Car & Battery Storage New Program Review, House of Lords, UK, July 2017.

UK Parliament, House of Commons Environmental Audit Committee, Green Finance Inquiry Oral Evidence Published Records, London UK, 20 February 2018.

UN, Greening the Economy with Agriculture Report, USA, 2012.

UN, United Nations Fact Sheet on Climate Change on Africa, USA, 2018.

UN Brundtland Commission, Report: Our Common Future, WCED UN, USA, 1987.

UN Environment, Global Emissions Gap Report, USA, 2017.

UN Habitat & MIT, Floating City to fight Climate Change, USA, 2019.

UN, United Nations 2015 Sustainable Development Goals (SDGs), NYC, USA, 2015.

UNEP, District Energy in Cities: Unlocking the Potential of Energy Efficiency and Renewables, USA, 2018.

UNFCCC, Distributed Renewable Power Generation and Integration, USA, 2015.

UNFCCC, Kyoto Protocol: Targets for the First Commitment Period, USA, 2012.

UNFCCC, Biennial Assessment and Overview of Climate, NYC, USA, 2016.

UNFCCC, United Nations Framework Convention on Climate Change, Paris Agreement, USA, 2016.

UN IPCC, Special Report CCS SRCCS, USA, 2005.

UN IPCC, Fourth Assessment Report on Climate Change, USA, 2007.

UN IPCC, Summary for Policymakers in Climate Change, Cambridge University Press, USA, 2007.

UN IPCC, Special Report on Renewable Energy Sources and Climate Change Mitigation, USA, 2012.

UN IPCC, Evaluation of Climate Models, USA 2013.

UN IPCC, Report on Climate Change, USA, 2017.

UN IPCC, IPCC Special Report on the Impacts of Global Warming, USA, 2018.

US EPA, Global Mitigation of Non-CO2 Greenhouse Gases, USA, 2012.

US EPA, Climate Change Science, Causes of Climate Change, USA, 2016.

US EPA, Overview of Greenhouse Gases in Greenhouse Gas (GHG) Emissions, USA, 2018.

US EPA, Global Greenhouse Gas Emission Database, USA, 2019.

US National Hurricane Centre, Tropical Hurricane Report, USA, 2018.

USA Oak Ridge National Lab, Boden, T.A., G. Marland, and R.J. Andres, 'Global, Regional, and National Fossil-Fuel CO2 Emissions', Carbon Dioxide Information Analysis Center, U.S. Department of Energy, Oak Ridge, Tenn., USA, 2013.

USA Office of the Deputy Assistant Secretary of the Army (Research & Technology), Emerging Science and Tech Trends, 2017–2047, USA, November 2017.

Wang, H., Canada Patent CA2008347, 'Removing Hydrogen Cyanide and Carbon Oxy-Sulphide from a Syngas Mixture', Shell Internationale, Canada, 1990.

Wang, H., University of Leeds, Visiting Lecturer on the 'Successful Chemical Plant Start-up & Commissioning' Course at the University of Leeds in UK from 1993 to 1996, Lectures on the 'Major Ethyl Benzene Chemical Plant Start-up & Commissioning at Shell UK Stanlow Refinery', University of Leeds, UK, 1993.

Wang, H. with Mobil USA & Raytheon USA Authors, *Oil & Gas Journal (OGJ)* Paper on UK Refinery Successful Demonstrations of New Ethyl Benzene Process, USA, 1995.

Wang, H., Imperial College of Science & Technology MSc DIC Thesis on Bubble Flow Biological Reactor Research & Developments, Imperial College London, UK, 1997.

Wang, H., Singapore Prime Minister Office high level China Strategy Meeting Speech & Presentation on China Social, Economic and Industrial developments & China Strategy Developments, Singapore, 2001.

Wang, H., China Ministry of Commerce & Foreign Trade MOFCOM Transnational Company Forum Speech on New Development Strategy of Transnational Companies in China, MOFCOM, Beijing, PRC, 2003.

Wang, H., China National Development Reform Commission NDRC Energy Research Institute ERI Report on China Medium & Long Term Energy & Carbon Scenarios Report in 2003 jointly by China Energy Research Institute of the PRC Government National Development Reform Commission with USA Lawrence Berkley Lab of USA Government Department of Energy & Shell Group Planning, NDRC ERI, Beijing, PRC, 2003.

Wang, H., *China Daily* 2003 news interview report in Beijing on EU Work Group Energy Proposals to Government with Henry Wang, Chairman of EU Energy, Petrochemicals, Oil & Gas Committee, *China Daily*, Beijing, PRC, 2003.

Wang, H., Wharton Shell Group Business Leadership Program Business Case paper, Wharton Business School, Pennsylvania, USA, 2004.

Wang, H., China SASAC Minister Meeting in Beijing China in 2004, speech & paper on China Energy Scenarios & Challenges, SASAC, Beijing, PRC, 2004.

Wang, H., UK Prime Minister Climate Change Adviser & DEFRA Director General Ministerial Meeting in London UK in 2004, presentation & paper on China Energy & Climate Change Outlooks, DEFRA, London, UK, 2004.

Wang, H., China Ministry of Foreign Affairs & Institute of International Cooperation Meeting in Beijing in 2004, speech & paper on China Energy Business Outlooks & International Co-operations, MOFCOM, Beijing, PRC, 2004.

Wang, H., USA & UK Counsel Generals Meetings in Shanghai China in 2004, presentation & paper on China Energy Outlooks, UK Embassy, Shanghai, PRC, 2004.

Wang, H., London School of Economics Lecture & paper on China Outlooks & Opportunities, London School of Economics LSE, UK, 2004.

Wang, H., China Economics Round Table in Beijing speech & paper on China Clean Energy Sustainable Developments, China Economics Roundtable, Beijing, PRC, 2004.

Wang, H., et al., UK China Bilateral Energy Strategic Cooperations Paper with UK China Bilateral Energy Work Group, UK Embassy, Beijing, PRC, 2005.

Wang, H., International Advanced Management Seminar papers on Global & China Business Issues, Energy Planning in China, Government structures in China & New Business Development in China, New York Bar Association HQ, New York, USA, 2005.

Wang, H., *China Daily* CEO Corporate Social Responsibilities Round Table in Beijing in 2005 interview on CSR, *China Daily*, Beijing, PRC, 2005.

Wang, H., China Global Economic & Leadership Summit, Speech on Energy Economic Developments by Henry Wang with USA Nobel Economists at the Grand Hyatt Hotel in Beijing organised by China Cajing Economic Publishing Group, Beijing, PRC, 2005.

Wang, H., China State Council Development Research Council [DRC] in Beijing in 2005 Presentation on Global & China Energy Scenarios, DRC Beijing, PRC, 2005.

Wang, H., China Ministry of Foreign & Economic Cooperation [MOFCOM] Summit in China in 2005, speech & paper on Multinational Co Co-operations & Sustainable Developments in China, MOFCOM, Beijing, PRC, 2005.

Wang, H., China Economic Summit at Great Hall of People in Beijing in 2005, paper on China Energy Outlooks & Scenarios by Henry Wang, China Cajing Economic Magazine, Beijing, PRC, 2005.

Wang, H., Netherlands Energy Minister Meeting in Beijing in 2005, Presentation on China Energy Developments, Netherlands Embassy, Beijing, PRC, 2005.

Wang, H., Tsinghua University Lecture in Beijing China in 2005, Lecture & paper on Multinational Cos Operations in China, Tsinghua University, Beijing, PRC, 2005.

Wang, H., Board Meeting of a leading International Chemical Company & a top Middle East Company Joint Venture in Singapore in 2005, presentation on China Economic & Energy Outlooks, Singapore, 2005.

Wang, H., China Energy & Strategy Seminar for a leading Asia Government Prime Minister Office PMO & key Ministries presentations & papers on China Energy Planning & Developments, Market Access & Cooperation Strategies & Challenges, Shell Asia, 2005.

Wang, H., *China Daily* CEO Climate Change Round Table in Beijing in 2006 interview on Climate Change Outlooks by Henry Wang, *China Daily*, Beijing, PRC, 2006.

Wang, H., Energy Seminar for PRC Government Top Officials at Joint Tsinghua Harvard MPA Course presentation & paper on Global Energy Planning, Advanced Technologies & Management to Vice Ministers/Governors, Tsinghua University, Beijing, PRC, 2006.

Wang, H., Climate Change & Sustainable Development Seminar for PRC Government Senior Officials presentation & paper on International Sustainable Development, Climate Change, Carbon technologies & management, Tsinghua University, Beijing, PRC, 2006.

Wang, H., China Advanced Management Seminar in Beijing China for top international executives, presentations & papers on China Business Issues, China Energy Planning & China Business Developments, Shell Beijing, PRC, 2006.

Wang, H., EU Chamber of Commerce China in Beijing in 2007 Presentation on Clean Energy Developments & Opportunities, EUCCC, Beijing, PRC, 2007.

Wang, H., China State Council Development Research Council Presentation on Clean Energy & Coal Developments in China & Globally, DRC Beijing, PRC, November 2008.

Wang, H., Remin University and China Carbon Forum Conference in 2008, speech & paper on Clean Coal Developments and Copenhagen Negotiations, Remin University, Beijing, PRC, 2008.

Wang, H., UCL Distinguished Speaker lecture on China Advanced Coal Technology & Successful Project Developments at University College London, London, UK, 2008.

Wang, H., China Netherlands Prime Ministerial Energy Summit at Tsinghua University in Beijing in Nov 2008, paper on Integrated Energy Management, Clean Energy & Sustainable Development, Tsinghua University, Beijing, PRC, 2008.

Wang, H., China International Radio interview in Beijing China in April 2010 on World Bank Six Asia Country Energy Report, China Radio, Beijing, PRC, 2009.

Wang, H., Argus Carbon Report Interview in London UK in Oct 2009 on China Carbon & Climate Change Trends by Henry Wang with Argus Carbon Editor, Argus, London, UK, 2009.

Wang, H., Bloomberg News interview in Singapore in Oct 2009 on China Climate Change Policies Outlooks by Bloomberg Asia Editor, Bloomberg, Singapore, 2009.

Wang, H., Carbon Forum Asia in Singapore in Oct 2009 keynote speech & Paper on China Climate Change & Sustainable Development Policies, Singapore, 2009.

Wang, H., Carbon Forum Asia in Singapore in Oct 2009 Paper on China Carbon Market Management & Outlooks, Singapore, 2009.

Wang, H., UK China Chemicals CEO Working Group Forum in Shanghai in Nov 2009, Paper & Presentation on Integrated Energy Management, Clean Energy Technologies & Sustainable Developments in China, UK Embassy, Shanghai, PRC, 2009.

Wang, H., UK Embassy China in Beijing in 2009 presentation on China Clean Energy & Sustainable Development by Henry Wang, UK Embassy, Beijing, PRC, 2009.

Wang, H., China Carbon Forum Government Round Table keynote speech on China Clean Energy & Carbon Developments, Remin University, Beijing, PRC, June 2009.

Wang, H., Speech & Presentation on Shale Gas Business Growth, Commercialisation & Developments in China, to the First China International Shale Gas Conference on 26–27 October 2010 in Shanghai organised by IBC Asia, China, 2010.

Wang, H., Deep-water Drilling Outlook Summit in Singapore in July 2010 Paper on China Upstream Offshore Developments, Singapore, 2010.

Wang, H., Asia Pacific Offshore Support Forum in Singapore in April 2010 Paper on China Offshore Support Industry Developments, Singapore, 2010.

Wang, H., China International & Beijing State Radio interview in Beijing on International Earth Day in April 2010 on Green Energy, Renewables, Chemicals, Coal Gasification, Energy Efficiency & Sustainable Developments, China Radio, Beijing, PRC, 2010.

Wang, H., Keynote speech to First International Four Kingdom Carbon International Conference organised by Saudi Ministry of Petroleum in Saudi Arabia, 2011.

Wang, H., China Market Developments & Marketing Lecture in April 2012 to EMBA class at University of Colorado Denver Business School, Denver, Colorado, USA, 2012.

Wang, H., International Energy & Renewables Strategic Co-Development Lecture in April 2012 at University of Colorado Denver Business School, Denver, Colorado, USA, 2012.

Wang, H., Global & Middle East Petrochemical Growth & Developments at University of Colorado Energy Conference, Boulder, Colorado, USA, 2012.

Wang, H., India Oil IOC Chairman Petrochemical Conclave presentation 'Opportunities & Challenges in Industries Winning Strategies', IOC, Delhi, India, March 2012.

Wang, H., International Energy Agency IEA Energy Efficiency EE Manual review commentary to OECD BIAC and IEA EE Team in August 2013, IEA, Paris, France, 2013.

Wang, H., Presentation to China Ministry of Commerce & China National Oil Companies Delegation visit to Saudi Arabia in May 2013, SABIC, Riyadh, Saudi Arabia, 2013.

Wang, H., International Energy Agency World Energy Outlook Peer Review Panel – Global Energy Competitiveness inputs to the IEA WEO Team in April 2013, IEA, Paris, France, 2013.

Wang, H., IEA, OPEC and IEF International Energy Conference presentation at IEF HQ in Riyadh Saudi Arabia, January 2013.

Wang, H., UK Chartered Management Institute Management Paper of Year 2014 Submission titled 'Business Negotiation Strategy & Planning in China', CMI London, UK, 2014.

Wang, H., International Energy Agency World Energy Outlook (IEA WEO) Peer Review Panel Global Energy & Petrochemical Investment Cost Reviews commentaries to the IEA WEO Team in IEA HQ in January 2014, IEA HQ, Paris, France, 2014.

Wang, H., *Successful Business Dealings and Management with China Oil, Gas and Chemical Giants*, Routledge, 2014.

Wang, H., Fourth International Energy Forum & International Energy Authority & OPEC Symposium on Energy Outlooks speech on 'Petrochemicals & Chemicals Growth Outlooks & Strategic Developments' in IEF HQ in Riyadh, Saudi Arabia, 22 January 2014.

Wang, H., Presentation on Sustainable Petrochemical & Chemicals Outlooks to OECD Energy & Environmental Committee Meetings, OECD Paris, France, 26 February 2014.

Wang, H., Presentation on Sustainable Petrochemical & Chemicals Outlooks to 2nd IEA Unconventional Gas Forum on 26 March 2014, Calgary, Canada, March 2014.

Wang, H., King Abdullah Petroleum Studies & Research Centre KAPSARC First International Seminar on China keynote speech & presentation on 'Sustainable Growth Scenarios & Strategies' KAPSARC HQ in Riyadh, Saudi Arabia, March 2014.

Wang, H., KAPSARC Paper on Energy Productivity Aligning Global Agenda Peer Review comments, KAPSARC HQ, Riyadh, Saudi Arabia, April 2014.

Wang, H., OECD BIAC China Task Force Presentation to OECD China Reflection Group & OECD Ambassadors consultation inputs, OECD BIAC, Paris, France, 23 & 24 June 2014.

Wang, H., Japan Ministry Economic Trade Industry METI Presentation on Saudi Arabia Downstream Industrial Cluster Development Program, SABIC, Riyadh, Saudi Arabia, June 2014.

Wang, H., UK CBI White Paper on Business Energy and Climate Change Priorities for the 2015–2020 UK Parliament consultation inputs, UK CBI, London, UK, August 2014.

Wang, H., OPEC IEA IEF Energy Conference & IEF KAPSARC Energy Roundtable 23–4 March 2015 discussion inputs, IEF Riyadh Saudi Arabia, March 2015.

Wang, H., ICIS 9th Asia Base Oil & Lubricant Conference keynote speech on 'China Demand Growth & Sustainable Growth Strategies', Singapore, 10 June 2015 in Singapore.

Wang, H., Singapore Energy Week Asia Downstream Conference keynote speech and presentation on 'Global Supply Chain Management, Risk Minimisation, Resource and Cost Optimisation Strategies', Singapore 28 October 2015.

Wang, H., UK Chartered Management Institute Top Five Management Paper of Year 2015, China Business Negotiation Strategy, CMI London, UK, February 2016.

Wang, H., OECD Integrity Forum paper, Global and MENA SOE Governance and Integrity, Paris, France, April 2016.

Wang, H., Transparency International SOE Integrity Forum paper, Global SOE Management and Governance Improvements, Berlin, Germany, June 2016.

Wang, H., EU Chamber of Commerce China Energy Panel, Energy Markets in Emerging Economies, Beijing, PRC, August 2016.

Wang, H., China Academy of Science Dalin Institute, Energy Growth Strategies, Dalin, China, November 2016.

Wang, Henry, *Energy Markets in Emerging Economies: Strategies for Growth*, Abingdon and New York: Routledge, 2017.

Wang H., Sino-British Summit Paper, China NDRC Renewables and Smart City Plans, UK, 2017.

Wang, H., Kings College London, International Energy and Environment Growth Strategies, London, UK, January 2017.

Wang, H., Imperial College London, Energy and Environment Growth Strategies, London, UK, February 2017.

Wang, H., Chinese University of Hong Kong, Energy, Environment and Climate Change, Hong Kong, March 2017.

Wang, H., Hong Kong University Speech, Energy, Environment and Climate Change Action Plans, HKU, Hong Kong, 9 April 2017.

Wang, H., Hong Kong Science Tech Association Speech, Energy, Environment and Climate Change Innovations, Hong Kong, 21 April 2017.

Wang, H., Liechtenstein International Economic Forum Speech, China Economic Growths, Liechtenstein, June 2017.

Wang, H., Imperial College Business School MSc. Climate Finance Lecture, Climate Change & Green Finance Growths, London UK 24 October 2017.

Wang, H., ICIS Global Base Oil Conference Speech, China Belt & Road Initative Growths, London, UK, 21 February 2018.

Wang, H., HK Rotary Taipo Club Speech, International & China Business Negotiations, Hong Kong, 14 May 2018.

Wang, H., HK Rotary Peninsula Club speech, Climate Change & Climate Action Plan HK, Hong Kong, May 2018.

Wang, H., University College London HK Alumni Association Speech, Business Negotiations, Hong Kong, June 2018.

Wang, H., India Institute of Director Global Convention Proceedings Paper, Climate Change & Climate Finance TCFD Reporting, Mumbai, India, July 2018.

Wang, H., Hong Kong Dragon Foundation Youth Leaders Speech, International & China Business Negotiations, Hong Kong City University, Hong Kong, 25 August 2018.

Wang, H., London School of Economics Negotiation Society Lecture, Business Negotiations in China, LSE, London, UK, 17 October 2018.

Wang, H., India Institute of Director Global Convention London Paper, Climate Change & Green Finance Governance Growths, London, UK, 25 October 2018.

Wang, H., FT Asia Climate Finance Summit, Renewables Growth, Challenges and Opportunities, Hong Kong, 21 November 2018.

Wang, H., UK House of Lords Energy Panel Paper, China Fossil & Renewables Energy Transformations, London, UK, 5 December 2018.

Wang, H., HKUST Post COP24 Forum 'Hong Kong Climate & Decarbonisation Challenges', Hong Kong, 21 January 2019.

Wang, H., HK City University Colloquium, China & Hong Kong Climate & Decarbonisation Challenges, Hong Kong, 21 February 2019.

Wang, H., UK House of Lords Westminister Energy Group Windsor Summit, China Climate & Energy Challenges, Windsor, UK, 2 March 2019.

Wang, H., Hong Kong Green Council Climate Forum, Global & Hong Kong Climate & Decarbonisation Challenges, Hong Kong, 9 May 2019.

Wang, H., HKUST Business School Skolkovo EMBA Lectures, Doing Business in China and Asia, Hong Kong, 10 July 2019.

Wang, H., Management Association Philippines CEO Conference Speech on Business Sustainability, Impacts and Future, Manila, 10 September 2019.

Wang, H., Oil and Money Conference Geopolitic Panel Speech, London, UK, 9 October 2019.

WECF, Fact Sheet Dangerous Health Effects of Home Burning of Plastics and Waste, USA, 2005.

Wiseman, E., Everything You Need to Know about the New UK Emission Rules, UK, July 2017.

WMO (World Meteorological Organisation), State of the Climate Report, Switzerland November 2018.

WMO (World Meteorological Organisation), State of the Climate Report, Switzerland, 2019.

World Bank, World Development Report 2010, USA, 2010.

World Bank, State and Trends of Carbon Pricing, USA, 2015.

World Bank, Developing East Asia Pacific Growth in 2015, USA, 13 April, 2015.

World Bank, Refugee Population by Country Data, USA, 2018.

World Bank, Groundswell: Preparing for Internal Climate Migration, World Bank Report, USA, March 2018.

World City Summit, Innovative Cities of Opportunity, Singapore, 2016.

World Economic Forum,The Global Competitiveness Report 2010–2011, Switzerland, 2010.

World Economic Forum's Fourth Industrial Revolution for the Earth series, How Technology is Leading Us to New Climate Change Solutions, Switzerland, 2018.

World Ocean Review, Climate Change and Methane Hydrates, USA, 2010.

World Science, Industrial Map of China Energy, USA, 2013.

WWEA (World Wind Energy Association), Small Wind World Report, Bonn, Germany, 2017.

WWF (World Wildlife Fund), Climate change, Coral Reefs and the Coral Triangle, USA, 2019.

Xiaowen Tian, *Managing International Businesses in China*, Cambridge University Press, UK, 2007.

Xinhua News Agency, PRC President Xi Jinping Joint Written Interview to the Media of Trinidad and Tobago, Costa Rica and Mexico on 31 May 2013, Xinhua News Agency, Beijing, PRC, 2013.

Xinhua News Agency, China M&A 2012 Highlights, 22 May 2013, Beijing, PRC, 2013.

Xinhua News Agency, Chinese Carbon Emissions to Peak in 2030, Beijing, PRC, 2014.

Xinhua News Agency, China to Reduce Coal Consumption for Better Air, Beijing, PRC, 2015.

Xinhua News Agency, China to Create Xiongan New Area in Hebei, Beijing, 2017.

Yale Environment, China Waste to Energy Incineration, USA, 2017.

Yuen, L., *Enterprising China*, Oxford University Press, UK, 2011.

Yuen, L., *China's Growth*, Oxford University Press, UK, 2013.

ZEV Alliance (International Zero Emission Vehicle Alliance), Zero Emissions Target Announcements, USA, 2015.

Zhang, X., Karplus, V.J., Qi, T., Zhang, D. & He, J., MIT Joint Report Series, Carbon Emissions in China: How Far Can New Efforts Bend the Curve?, MIT, USA, 2014.

Glossary

AAU	Assigned Amount Unit
ADAS	Advanced Driver-Assistance Systems
AEV	Autonomous Electric Vehicle
AI	Artificial Intelligence
APF	Antarctic Polar Front
BECCS	Bio-energy Carbon Capture and Storage
BECCU	Bio-energy Carbon Capture Utilisation
BEV	Battery Electric Vehicle
BP	British Petroleum
CAPEX	Capital Expenditure
CCS	Carbon Capture and Storage
CCU	Carbon Capture and Utilisation
CDM	Clean Development Mechanism
CEO	Chief Executive Officer
CER	Certified Emission Reduction
CERC	Comparative Education Research Centre
CETS	Carbon Emission Trading Scheme
CFCs	Chlorofluorocarbons
CM	Cubic Meters
CMI	Confederation of Management Institute
CMPY	Cubic Meters per Year
CNG	Compressed Natural Gas
CNPC	China National Petroleum Corporation
CO$_2$	Carbon Dioxide
COP	Conference of Parties
CSO	Combined Sewer Overflows
CSR	Corporate Social Responsibility
DDT	Dichloro-Diphenyl-Trichoroethane
DICP	Dalian Institute of Chemical Physics
DME	Dimethyl Ether
DNL	Dalian National Laboratory for Clean Energy
ECBM	Enhanced Coalbed Methane
EGR	Exhaust Gas Recirculation

EIA	Environment Impact Assessment
EOR	Enhanced Oil Recovery
EPA	Environmental Protection Agency
EPB	Environmental Protection Bureau
ESG	Environment, Social, Governance
ETS	Emissions Trading Scheme
EUA	EU Allowance
EV	Electric Vehicle
FDI	Foreign Direct Investment
GDP	Gross Domestic Product
GFI	Green Finance Initiative
GHG	Greenhouse Gas
GM	Genetically Modified
Gt/y	Gigatonnes per year
GW	Gigawatts
GWP	Global Warming Potential of GHGs
H2	Hydrogen
HKEX	Hong Kong Stock Exchange
ICE	Internal Combustion Engines
ICT	Internet Computer Technology
IEA	International Energy Agency
IFES	Integrated Food Energy Systems
IMF	International Monetary Fund
INRM	Integrated Natural Resource Management
IOC	International Oil Company
IOT	Internet of Things
IPCC	Intergovernmental Panel on Climate Change
IPM	Integrated Pest Management
LDC	Least Developed Country
LDM	Long Distance Migrant
LEN	Low Emission Neighbourhood
LNG	Liquefied Natural Gas
LPG	Liquefied Petroleum Gas
LULUCF	Land Use, Land-use Change and Forestry
M&A	Mergers and Acquisitions
MENA	Middle East and North Africa
MOF	Ministry of Finance
MOFCOM	Ministry of Commerce of China
Mt	Metric Tonnes
MTO/MTP	Methanol to propylene and olefin
NCE	New Climate Economy
NDC	Nationally Determined Contributions
NDRC	National Development Reform Commission of China
NGO	Non-Governmental Organisation
NO/N20/NOX	Nitrogen Oxides

NPM	Non-pesticidal Management
NSIDC	National Snow & Ice Data Centre
OECD	Organisation for Economic Co-operation and Development
OEMS	Original Equipment Manufacturers
OOIP	Original Oil in Place
OPEC	Organisation of Petroleum Exporting Countries
OPEX	Operating Expenditure
PHEV	Plug-in Hybrid Electric Vehicle
PE	Poly-ethylene
PP	Polypropylene
PPM	Parts per Million
PRC	People's Republic of China
PV	Photovoltaics
R&D	Research & Development
REDD	Reducing Emissions from Deforestation and Forest Degradation
RFS	Renewable Fuel Standard
RGGI	Regional Greenhouse Gas Initiative
RMU	Ring Main Unit
SABIC	Saudi Arabia Basic Industries Co.
SASAC	State Assets Supervision and Administration Commission
SAT	Soil Aquifer Treatment
SECURE	Socio-economic and Cultural Upliftment in Rural Environment (India)
SF6	Sulfur Hexafluoride
SOE	State Owned Enterprise
SRI	Systems of Rice Intensification
tCO2e	Tonnes Carbon Dioxide Equivalent
TCFD	Task Force on Climate Finance Disclosure
TMG ETS	Tokyo Metropolitan Government Emission Trading System
TWh	Terrawatt Hour
UNFCCC	United Nations Framework Convention on Climate Change
US EPA	United States Environmental Protection Agency
WCI	Western Climate Initiative
WEO	World Energy Outlooks
WHO	World Health Organisation
WWF	World Wild Life Fund
ZEV	Zero-emission Vehicle

Index

Printed and bound by CPI Group (UK) Ltd, Croydon, CR0 4YY

22/10/2024

01777596-0006